AF361870

Symmetries in Quantum Mechanics and Statistical Physics

Symmetries in Quantum Mechanics and Statistical Physics

Editor

Georg Junker

MDPI • Basel • Beijing • Wuhan • Barcelona • Belgrade • Manchester • Tokyo • Cluj • Tianjin

Editor
Georg Junker
Institute for Theoretical Physics I
University Erlangen-Nürnberg
Erlangen
Germany

Editorial Office
MDPI
St. Alban-Anlage 66
4052 Basel, Switzerland

This is a reprint of articles from the Special Issue published online in the open access journal *Symmetry* (ISSN 2073-8994) (available at: www.mdpi.com/journal/symmetry/special_issues/symmetries_quantum_mechanics_statistical_physics).

For citation purposes, cite each article independently as indicated on the article page online and as indicated below:

LastName, A.A.; LastName, B.B.; LastName, C.C. Article Title. *Journal Name* **Year**, *Volume Number*, Page Range.

ISBN 978-3-0365-2479-5 (Hbk)
ISBN 978-3-0365-2478-8 (PDF)

Contents

About the Editor

Georg Junker

Georg Junker received his Ph.D. degree from Würzburg University (Germany) in 1989; subsequently, he assumed an assistant position at the University of Erlangen-Nürnberg (Germany), where he also received his Habilitation and became a Privatdozent. He acts as referee and advisory board member for several scientific journals and has authored more than 60 publication, including two books. He is currently affiliated with the European Southern Observatory in Garching (Germany) and the Institute for Theoretical Physics at the University of Erlangen-Nürnberg (Germany).

His research activities focus on all kinds of symmetry aspects of physical systems. He has made major contributions within Feynman's path-integral approach to quantum mechanics, and has also successfully applied group theoretical methods to problems in quantum chaos and statistical physics. His current research interest is on supersymmetric quantum mechanics and its application to quantum mechanics and statistical and solid-state physics.

Editorial

Special Issue: "Symmetries in Quantum Mechanics and Statistical Physics"

Georg Junker [1,2]

1   European Southern Observatory, Karl-Schwarzschild-Straße 2, D-85748 Garching, Germany; gjunker@eso.org
2   Institut für Theoretische Physik I, Universität Erlangen-Nürnberg, Staudtstraße 7,
     D-91058 Erlangen, Germany

Citation: Junker, G. Special Issue: "Symmetries in Quantum Mechanics and Statistical Physics". *Symmetry* **2021**, *13*, 2027. https://doi.org/10.3390/sym13112027

Received: 20 October 2021
Accepted: 21 October 2021
Published: 26 October 2021

Symmetry is a fundamental concept in science and has played a significant role since the early days of quantum physics. In physics, symmetry characterises the invariance of a system under certain transformations, being either discrete like mirror symmetry or continuous like rotational symmetry. In mathematics, symmetries are described by group theoretic means.

Symmetry methods are still powerful tools in contemporary problems of quantum mechanics and statistical physics and go beyond the classical Lie groups and algebras. Examples are the so-called supersymmetric quantum mechanics and the PT–invariance of non-Hermitian Hamiltonians, but also include duality concepts, besides others. This Special Issue presents recent contributions to such new fundamental symmetry concepts.

The work by Znojil [1] investigates non-Hermitian PT-symmetric extensions of Bose–Hubbard-like models. Particular focus is made on perturbations near so-called exceptional points, that is, points within the real spectrum of non-Hermitian Hamiltonians exhibiting degeneracy, and its stability under perturbations.

This special issue also contributes several new results on various topics related to supersymmetric quantum mechanics (SUSY QM). The work by Quesne [2] increases the number of exactly solvable quantum models by extending the shape-invariance concept of SUSY QM to deformed SUSY QM models. Gadella et al. [3] present a thorough and complete study of the various supersymmetric partners of the one-dimensional infinite square-well model, with particular focus on self-adjoint extensions. The contributions by the editor discuss the SUSY QM structure of relativistic Hamiltonians. In [4], generic relativistic Hamiltonians for an arbitrary spin are considered, and a SUSY QM structure is obtained under certain conditions. In a second contribution [5], the Klein–Gordon oscillator is discussed explicitly, and SUSY is utilised to find a closed form expression for the eigenvalues and eigenfunction, as well as for the corresponding Green's function.

The contributions by Inomata et al. [6] and Zhao et al. [7] reconsider joint transformations of space and time, mapping different physical systems onto each other. In [6], the well-known Newton–Hooke duality and its generalization to arbitrary power-law potentials is reviewed. Here, duality is viewed as a symmetry concept. The contribution [7] reconsiders space–time transformations, mapping a quadradic system onto that of a free particle. The close relation of this transformation with the time-dependent Bargmann-conformal transformation is established and illustrated.

Finally, the contribution by Schulman [8] takes a fresh look into the spreading of wave packets. Whereas wave packets spreading in a Gaussian-like manner may be localized due to scattering, it is argued that this localization may not happen for wave packets spreading faster than in a Gaussian manner.

References

1. Znojil, M. Perturbation theory near degenerate exceptional points. *Symmetry* **2020**, *12*, 1309. [CrossRef]
2. Quesne, C. Deformed shape invariant superpotentials in quantum mechanics and expansions in powers of $\hbar$. *Symmetry* **2020**, *12*, 1853. [CrossRef]
3. Gadella, M.; Hernandez-Munoz, J.; Nieto, L.M.; San Millan, C. Supersymmetric partners of the one-dimensional infinite square well Hamiltonian. *Symmetry* **2021**, *13*, 350. [CrossRef]
4. Junker, G. Supersymmetry of relativistic Hamiltonians for arbitrary spin. *Symmetry* **2020**, *12*, 1590. [CrossRef]
5. Junker, G. On the supersymmetry of the Klein–Gordon oscillator. *Symmetry* **2021**, *13*, 835. [CrossRef]
6. Inomata, A.; Junker, G. Power law duality in classical and quantum mechanics. *Symmetry* **2021**, *13*, 409. [CrossRef]
7. Zhao, Q.; Zhang, P.; Horvathy, P.A. Time-dependent conformal transformations and the propagator for quadratic systems. *Symmetry* **2021**, *13*, 1866. [CrossRef]
8. Schulman, L. What is the size and shape of a wave packet? *Symmetry* **2021**, *13*, 527. [CrossRef]

Article

Perturbation Theory Near Degenerate Exceptional Points

Miloslav Znojil  [1,2]

1   The Czech Academy of Sciences, Nuclear Physics Institute, Hlavní 130, 25068 Řež, Czech Republic;
    znojil@ujf.cas.cz
2   Department of Physics, Faculty of Science, University of Hradec Králové, Rokitanského 62,
    50003 Hradec Králové, Czech Republic

Received: 13 July 2020; Accepted: 4 August 2020; Published: 5 August 2020

Abstract: In an overall framework of quantum mechanics of unitary systems a rather sophisticated new version of perturbation theory is developed and described. The motivation of such an extension of the list of the currently available perturbation-approximation recipes was four-fold: (1) its need results from the quick growth of interest in quantum systems exhibiting parity-time symmetry ($\mathcal{PT}$-symmetry) and its generalizations; (2) in the context of physics, the necessity of a thorough update of perturbation theory became clear immediately after the identification of a class of quantum phase transitions with the non-Hermitian spectral degeneracies at the Kato's exceptional points (EP); (3) in the dedicated literature, the EPs are only being studied in the special scenarios characterized by the spectral geometric multiplicity L equal to one; (4) apparently, one of the decisive reasons may be seen in the complicated nature of mathematics behind the $L \geq 2$ constructions. In our present paper we show how to overcome the latter, purely technical obstacle. The temporarily forgotten class of the $L > 1$ models is shown accessible to a feasible perturbation-approximation analysis. In particular, an emergence of a counterintuitive connection between the value of L, the structure of the matrix elements of perturbations, and the possible loss of the stability and unitarity of the processes of the unfolding of the singularities is given a detailed explanation.

Keywords: non-Hermitian quantum dynamics; unitary vicinity of exceptional points; degenerate perturbation theory; Hilbert-space geometry near EPs

---

1. Introduction

The Bender's and Boettcher's [1] idea of replacement of Hermiticity $H = H^\dagger$ by parity-time symmetry ($\mathcal{PT}$-symmetry) $H\mathcal{PT} = \mathcal{PT}H$ of a Hamiltonian responsible for unitary evolution opened, after an appropriate mathematical completion [2–5] of the theory, a way towards the building of quantum models exhibiting non-Hermitian degeneracies [6] *alias* exceptional points (EPs, [7]). For a fairly realistic illustrative example of possible applications opening multiple new horizons in phenomenology we could recall, e.g., the well-known phenomenon of Bose–Einstein condensation is a schematic simplification described by the non-Hermitian but $\mathcal{PT}$-symmetric three-parametric Hamiltonian

$$H^{(GGKN)}(\gamma, v, c) = -\mathrm{i}\gamma \left(a_1^\dagger a_1 - a_2^\dagger a_2 \right) + v \left(a_1^\dagger a_2 + a_2^\dagger a_1 \right) + \frac{c}{2} \left(a_1^\dagger a_1 - a_2^\dagger a_2 \right)^2 . \tag{1}$$

This Hamiltonian represents an interesting analytic-continuation modification of the conventional Hermitian Bose–Hubbard Hamiltonian [8–10]. In this form the model was recently paid detailed attention in Ref. [11]. A consequent application of multiple, often fairly sophisticated forms of perturbation theory has been shown there to lead to surprising results. In particular, the behavior of the bound and resonant states of the system was found to lead to the new and unexpected phenomena in the dynamical regime characterized by the small coupling constant c. The authors of Ref. [11] emphasized that new physics may be expected to emerge precisely in the vicinity of the EP-related dynamical singularities.

These phenomena (simulating not necessarily just the Bose–Einstein condensation of course) were analyzed, in [11], using several ad hoc, not entirely standard perturbation techniques. The role of an unperturbed Hamiltonian H_0 was assigned, typically, to the extreme EP limits of H. Unfortunately, only too often the perturbed energies appeared to be complex as a consequence. In other words, the systems exhibiting $\mathcal{PT}$-symmetry seemed to favor the spontaneous breakdown of this symmetry near EPs.

In the light of similar results one immediately must ask the question whether such a "wild behavior" of the EP-related quantum systems is generic. Indeed, an affirmative answer is often encountered in the studies by mathematicians (see, e.g., [12]). A hidden reason is that they usually tacitly keep in mind just the "effective theory" and/or the so-called "open quantum system" dynamical scenario [13].

In the more restrictive context of the unitary quantum mechanics the situation is different: the "wild behavior" of systems is usually not generic there (cf. also the recent explanatory commentary on the sources of possible misunderstandings in [14]). Several non-numerical illustrative models may be found in [15] where typically, the $\mathcal{PT}$-symmetric Bose–Hubbard model of Equation (1) (which behaves as unstable near its EP singularities [11])) has been replaced by its "softly perturbed" alternative in which in an arbitrarily small vicinity of its EP singularity, the system remains stable and unitary under admissible perturbations.

The physics of stability covered by paper [15] can be perceived as one of the main sources of inspiration of our present study. We intend to replace here the very specific model of Equation (1) (in which the geometric multiplicity L of all of its EP-related degeneracies was always equal to one) by a broader class of quantum systems. In a way motivated by the idea of a highly desirable extension of the currently available menu of the tractable and eligible dynamical scenarios beyond their $L = 1$ subclass, we will turn attention here to the EP-related degeneracies of the larger, nontrivial geometric multiplicities $L \geq 2$. We will reveal that such a study opens new horizons not only in phenomenology (where the influence of perturbations becomes strongly dependent on the detailed structure of the non-Hermitian degeneracy) but also in mathematics (where a rich menu of physical consequences will be shown reflected by an unexpected adaptability of the geometry of the Hilbert space to the detailed structure of the perturbation).

The presentation of our results will be organized as follows. First, in Section 2 we will recall a typical quantum system (viz., a version of the non-Hermitian Bose–Hubbard multi-bosonic model) in which the EP degeneracies play a decisive phenomenological role. We will explain that although the model itself only exhibits the maximal-order $L = 1$ EP degeneracies, such an option represents, from the purely formal point of view, just one of the eligible dynamical scenarios. In Appendix A a full classification of the EPs is presented therefore, showing, i.a., that the number of the "anomalous" EPs of our present interest with $L \geq 2$ exhibits an almost exponential growth at the larger matrix dimension N.

The goals of our considerations are subsequently explained in section 3. For the sake of brevity, we just pick up the first nontrivial case with $L = 2$, and we emphasize that even in such a case the basic features of an appropriate adaptation of perturbation theory may be explained, exhibiting also, not quite expectedly, a survival of the fairly user-friendly mathematical structure.

In order to make our message self-contained, the known form of the EP-related perturbation formalism restricted to $L = 1$ is reviewed in Appendix B. On this background, in a way based on a not too dissimilar constructive strategy, our present main $L \geq 2$ results are then presented and described in Section 4. We emphasize there the existence of the phenomenological as well as mathematical subtleties of the large-L models. We show that in our generalized, degenerate-perturbation-theory formalism a key role is played by an interplay between its formal mathematical background (viz, the non-Hermiticity of the Hamiltonians) and its phenomenological aspects (typically, the knowledge of L must be complemented by an explicit knowledge of the partitioning of Schrödinger equation).

In Section 5, all these aspects of the $L > 1$ perturbation theory are summarized and illustrated by a detailed description of the characteristic, not always expected features of the leading-order approximations. Several related applicability aspects of our present degenerate-perturbation-theory formalism are finally discussed in Section 6 and in two Appendices. We point out there that some of the features of the $L > 1$ theory (e.g., a qualitative, fairly counterintuitive clarification of the concept of the smallness of perturbations) may be treated as not too different from their $L = 1$ predecessors. At the same time, a wealth of new formal challenges is emphasized to emerge, in particular, in the analyses of the role of perturbations in the specific quantum systems which are required unitary.

2. Exceptional Points

2.1. Bose–Hubbard Model and Exceptional Points of Geometric Multiplicity One, $L = 1$

Illustrative $\mathcal{PT}$-symmetric Bose–Hubbard Hamiltonian operator (1) commutes with the number operator

$$\widehat{N} = a_1^\dagger a_1 + a_2^\dagger a_2 . \tag{2}$$

This means that the number of bosons N is conserved so that after its choice the Hamiltonian may be represented by a finite-dimensional non-Hermitian K by K matrix $H^{(GGKN)}(K, \gamma, v, c)$ with $K = N + 1$ (see its explicit construction in [11]). Once we fix the units (such that $v = 1$) and once we set $c = 0$ (preserving, for the sake of simplicity, just the first two components of the Hamiltonian), the resulting one-parametric family of Hamiltonian matrices $H^{(GGKN)}(K, \gamma)$ can be assigned the closed-form energy spectra

$$E_n^{(GGKN)}(K, \gamma) = (1 - \gamma^2)^{1/2} (1 - K + 2n), \quad n = 0, 1, \ldots, K - 1 . \tag{3}$$

These energies remain real and non-degenerate (i.e., observable) if and only if $\gamma^2 < 1$.

In *loc. cit.* it has also been proved that the two interval-boundary values of $\gamma = \pm 1$ are, in the terminology of the Kato's mathematical monograph [7], exceptional points (EPs, $\gamma_+^{(EP)} = 1$ and $\gamma_-^{(EP)} = -1$). More precisely, one should speak about the very special EPs of maximal order (i.e., of order K, abbreviated as EPK). The latter observation may be given a more general, model-independent linear-algebraic background via relation

$$H^{(K)}(\gamma^{(EPK)}) Q^{(EPK)} = Q^{(EPK)} J^{(K)}(\eta) \tag{4}$$

where

$$\eta = \lim_{\gamma \to \gamma^{(EPK)}} E_n(K, \gamma), \quad n = 0, 1, \ldots, K - 1 \tag{5}$$

is the limiting degenerate energy. Relation (4) contains the so-called transition matrix $Q^{(EPK)}$ and the non-diagonal, EPK-related canonical-representation Jordan-block matrix

$$J^{(K)}(\eta) = \begin{pmatrix} \eta & 1 & 0 & \cdots & 0 \\ 0 & \eta & 1 & \ddots & \vdots \\ 0 & 0 & \eta & \ddots & 0 \\ \vdots & \ddots & \ddots & \ddots & 1 \\ 0 & \cdots & 0 & 0 & \eta \end{pmatrix} . \tag{6}$$

As a certain limiting analogue of the conventional set of eigenvectors the transition matrix is obtainable via the solution of the EPK-related analogue (4) of conventional Schrödinger equation. For our illustrative example $H^{(GGKN)}(K, \gamma)$, in particular, all of the K- and $\gamma^{(EPK)}$-dependent explicit, closed forms of solutions $Q^{(EPK)}$ remain non-numerical and may be found constructed in dedicated paper [15].

2.2. Generic Non-Hermitian Degeneracies with Geometric Multiplicities Larger than One, $L > 1$

In the common model-building scenarios the γ-dependence of Hamiltonians $H^{(K)}(\gamma)$ is analytic. Under this assumption the exceptional points of maximal order as discussed in preceding subsection represent just one of several possible realizations of a non-Hermitian degeneracy (NHD) with its characteristic EP-related confluence of eigenvalues (5). Besides the maximal, EPK-related complete confluence of eigenvectors as described in preceding subsection we may encounter, in general, multiple other, incomplete confluences of eigenvectors

$$\lim_{\gamma \to \gamma^{(NHD)}} |\psi_{m_{k[j]}}(\gamma)\rangle = |\chi_j\rangle\,, \quad k[j] = 1, 2, \ldots, M_j\,, \quad M_j \geq 2\,, \quad j = 1, 2, \ldots, L\,. \tag{7}$$

Here we have $M_1 + M_2 + \ldots + M_L = K$ where L is called the geometric multiplicity of the EP degeneracy [7] (see also Appendix A for more details).

Every EP instant $\gamma^{(NHD)}$ may be characterized not only by the overall Hilbert-space dimension K and by the number $L \geq 1$ of linearly independent η-related states $|\chi_j\rangle$ of Equation (7) but also by a suitably ordered L-plet of the related subspace dimensions M_j. Thus, in the present $L \geq 2$ extension of preceding subsection the fully non-diagonal Jordan block of Equation (6) must be replaced by the more general block-diagonal canonical representation of the Hamiltonian,

$$\mathcal{J}^{(K)}(\eta) = \begin{pmatrix} J^{(M_1)}(\eta) & 0 & \cdots & 0 \\ 0 & J^{(M_2)}(\eta) & \ddots & \vdots \\ \vdots & \ddots & \ddots & 0 \\ 0 & \cdots & 0 & J^{(M_L)}(\eta) \end{pmatrix} = \bigoplus_{j=1}^{L} J^{(M_j)}(\eta)\,. \tag{8}$$

In parallel, EPK relation (4) must be replaced by its generalization

$$H^{(K)}(\gamma^{(NHD)})\, Q^{(NHD)} = Q^{(NHD)}\, \mathcal{J}^{(K)}(\eta)\,. \tag{9}$$

Naturally, the direct-sum structure of $\mathcal{J}^{(K)}(\eta)$ becomes reflected by a partitioned-matrix structure of transition matrices $Q^{(NHD)}$ which are, in general, not block-diagonal of course.

3. Unitary Processes of Collapse at $L = 2$

A priori one may expect that the existence of anisotropy of the Hilbert space as realized, in Equation (A18) at $L = 1$, by an elementary rescaling $B(\lambda)$ of the basis will also exist at any larger geometric multiplicity $L > 1$, i.e., in the unitary quantum systems with the more complicated structure of the NHD limiting *alias* quantum phase transition.

3.1. Quantum Physics behind "Degenerate Degeneracies" with $L = 2$

Hypothetically, the unitary evolution of any $\mathcal{PT}$-symmetric quantum system moving towards a hiddenly Hermitian EP degeneracy with geometric multiplicity two can be perceived as generated by a suitable diagonalizable Hamiltonian $H^{(K)}(\gamma)$ with real spectrum [3]. What is only necessary is that its (perhaps, properly renumbered) eigenvectors $|\Phi_j(\gamma)\rangle$ obey the $L = 2$ EP-degeneracy rule

$$\lim_{\gamma \to \gamma^{(EP)}} |\Phi_m(\gamma)\rangle = |\chi_a\rangle\,, \quad m = 0, 1, \ldots, M - 1\,, \tag{10}$$

$$\lim_{\gamma \to \gamma^{(EP)}} |\Phi_{M+n}(\gamma)\rangle = |\chi_b\rangle, \qquad n = 0, 1, \ldots, N-1. \tag{11}$$

The two limiting eigenvectors $|\chi_a\rangle$ and $|\chi_b\rangle$ are, by our assumption, linearly independent so that partitioned matrix (8), i.e., at $L = 2$ and $\eta = 0$, matrix

$$\mathcal{J}^{(M \oplus N)}(0) = \left[\begin{array}{cc} J^{(M)}(0) & 0 \\ 0 & J^{(N)}(0) \end{array} \right] \tag{12}$$

will represent the canonical form of our Hamiltonian in the EP limit. The corresponding transition matrix $Q^{(M \oplus N)}$ can be then obtained by the solution of the limiting version

$$H^{(K)}(\gamma^{(EP)}) \, Q^{(M \oplus N)} = Q^{(M \oplus N)} \, \mathcal{J}^{(M \oplus N)}(0) \tag{13}$$

of the initial Schrödinger equation. A few exactly solvable samples of the latter $L = 2$ degeneracy process $\gamma \to \gamma^{(EP)}$ can be found, e.g., in our recent paper [16].

In all the similar dynamical scenarios one can always find parallels with their simpler $L = 1$ predecessors. In particular, a return to the situation before the collapse as sampled, at $L = 1$, by Equation (A4) below, can be also given the following analogous form

$$\left[Q^{((M \oplus N)} \right]^{-1} H^{(K)}(\gamma) \, Q^{(M \oplus N)} = J^{(M \oplus N)}(0) + \lambda \, V^{(K)}(\gamma), \quad \gamma \neq \gamma^{(EP)} \tag{14}$$

of a "unitarity-compatible" perturbation-theoretic reinterpretation in which the perturbation $\lambda \, V^{(K)}(\gamma)$ is fully determined by the input matrix Hamiltonian $H^{(K)}(\gamma)$.

3.2. Unfoldings of Degeneracies under Random Perturbations at $L = 2$

In a close parallel to the $L = 1$ Schrödinger's bound-state problem (A6) let us now start the study of its $L > 1$ generalizations by considering the first nontrivial choice of degeneracy with the geometric multiplicity $L = 2$. In our present notation the corresponding Schrödinger equation then reads

$$\left[\mathcal{J}^{(M \oplus N)}(0) + \lambda \, V^{(K)} \right] |\Psi\rangle = \varepsilon |\Psi\rangle. \tag{15}$$

Without any loss of generality, we set again $\eta = 0$. After such a choice all the eigenvalues $\varepsilon = \varepsilon(\lambda)$ will remain small, and they will vanish in the formal unperturbed-system limit $\lambda \to 0$.

In a way paralleling Equation (12), any given matrix of perturbations has to be partitioned as well,

$$V^{(K)} = \left[\begin{array}{cc} V^{(M,M)} & V^{(M,N)} \\ V^{(N,M)} & V^{(N,N)} \end{array} \right]. \tag{16}$$

Since the four submatrices have dimensions indicated by the superscripts, we shall assume, at the beginning at least, that all the individual matrix elements of the perturbation matrix $V^{(K)}$ remain bounded at small λ. This means that the Hamiltonian is dominated by its unperturbed part (12) so that also the whole perturbed $L = 2$ Schrödinger Equation (15) has to be partitioned. In order to simplify the notation we shall write

$$|\Psi\rangle = \left(\begin{array}{c} |\psi^{(a)}\rangle \\ |\psi^{(b)}\rangle \end{array} \right), \quad |\psi^{(a)}\rangle = \left(\begin{array}{c} \psi_1^{(a)} \\ \psi_2^{(a)} \\ \vdots \\ \psi_M^{(a)} \end{array} \right), \quad |\psi^{(b)}\rangle = \left(\begin{array}{c} \psi_1^{(b)} \\ \psi_2^{(b)} \\ \vdots \\ \psi_N^{(b)} \end{array} \right) \tag{17}$$

using the curly-ket symbols for subvectors. This will enable us to proceed in a partial parallel with the widely studied non-degenerate-EP cases where one has $L = 1$ (see Refs. [17,18] or Appendix B below for a compact review).

4. Perturbation Theory at $L = 2$

4.1. The Recent Change of the Unitary-Evolution Paradigm

From the historical point of view the use of EPs in physics has not been immediate. Only during the last circa 20 years one notices a perceivable increase of the relevance of the concept in various branches of theoretical as well as experimental physics. Various innovative EP applications emerged ranging from the analyses of resonances in classical mechanics [19] and of the so called non-Hermitian degeneracies in classical optics [6] up to the studies of the wealth of phenomena in quantum physics of open quantum systems [13,20] or, finally, even of the closed, stable and unitarily evolving quantum systems [1,2].

All these developments contributed to the motivation of our present study. For the sake of definiteness, we restricted our attention to the framework of quantum physics, unitary or non-unitary. In this setting the traditional role of the EPs $\gamma^{(EP)}$ has always been two-fold. First, in the context of mathematics, the conventional analyticity assumptions about Hamiltonians

$$H(\gamma) = H(\gamma_0) + (\gamma - \gamma_0) H^{(1)} + (\gamma - \gamma_0)^2 H^{(2)} + \dots, \tag{18}$$

and the conventional power-series ansatz for energies

$$E_n(\gamma) = E_n(\gamma_0) + (\gamma - \gamma_0) E_n^{(1)} + (\gamma - \gamma_0)^2 E_n^{(2)} + \dots \tag{19}$$

(etc.) gave birth to the so-called Rayleigh–Schrödinger perturbation-expansion constructions of the Schrödinger-equation solutions. It has been revealed that the radius of convergence R of these perturbation-series solutions is determined by the position of the nearest EP in the complex plane of the parameter, $R = \min |\gamma_0 - \gamma^{(EP)}|$ [7].

In the other, direct applications of EPs, the localization of singularities $\gamma^{(EP)}$ only played an important traditional role in non-unitary, open quantum systems [13]. An explanation is easy: for any self-adjoint Hamiltonian $H(\gamma)$ characterizing a closed quantum system, the necessary reality of the parameter ($\operatorname{Im} \gamma = 0$) cannot be made compatible with the fact that *all* of the values of the eligible (i.e., not accumulation-point) EP parameters are complex, $\operatorname{Im} \gamma^{(EP)} \neq 0$.

The traditional paradigm has only been changed recently, after Bender with Boettcher [1] managed to turn attention of physicists' community to the existence of a broad class of Hamiltonians $H(\gamma)$ which happen to be non-Hermitian but parity-time symmetric ($\mathcal{PT}$-symmetric) in $\mathcal{K}$. One of the characteristic mathematical features of these Hamiltonians is that despite their non-Hermiticity, their *whole* spectrum $\{E_n\}$ may remain strictly real in a suitable *real* interval $\mathcal{D}$ of the unitarity-compatible parameters γ (see, e.g., monograph [4] for more details).

4.2. Rearrangement of Schrödinger Equation

With the two subscripts $j_{(a)}$ and $j_{(b)}$ running, in the curly-ket subvectors in (17), from 1 to $N_{(a)} = M$ and $N_{(b)} = N$, respectively, we will now only partially fix the norm by setting $\psi_1^{(a)} = \omega_{(a)}$ and $\psi_1^{(b)} = \omega_{(b)}$ or, in a self-explanatory shorthand, $\psi_1^{(a,b)} = \omega_{(a,b)}$. Next, a parallel to the $L = 1$ redefinition (A8) of wave functions will be found in its $L = 2$ extension

$$|\vec{y}^{(a,b)}\succ = \begin{pmatrix} y_1^{(a,b)} \\ y_2^{(a,b)} \\ \vdots \\ y_{N_{(a,b)}-1}^{(a,b)} \\ y_{N_{(a,b)}}^{(a,b)} \end{pmatrix} = \begin{pmatrix} \psi_2^{(a,b)} \\ \psi_3^{(a,b)} \\ \vdots \\ \psi_{N_{(a,b)}}^{(a,b)} \\ \Omega_{(a,b)} \end{pmatrix} \tag{20}$$

where the two new, temporarily variable elements $\Omega_{(a)}$ and $\Omega_{(b)}$ will have to be determined later. In terms of the four auxiliary symbols

$$|e^{(a,b)}\succ = \begin{pmatrix} 1 \\ 0 \\ \vdots \\ 0 \end{pmatrix}, \quad \Pi^{(a,b)} = \begin{bmatrix} 0 & 0 & 0 & \dots & 0 \\ 1 & 0 & 0 & \ddots & \vdots \\ 0 & \ddots & \ddots & \ddots & 0 \\ \vdots & \ddots & 1 & 0 & 0 \\ 0 & \dots & 0 & 1 & 0 \end{bmatrix} \tag{21}$$

of dimensions $N_{(a)} = M$ and $N_{(b)} = N$ we will further decompose

$$|\psi^{(a,b)}\succ = |e^{(a,b)}\succ \omega_{(a,b)} + \Pi^{(a,b)} |\vec{y}^{(a,b)}\succ \ .$$

In the next step we introduce the $L = 2$ analogue of the unpartitioned $L = 1$ vector (A7),

$$|r\rangle = \begin{pmatrix} |r^{(a)}\succ \\ |r^{(b)}\succ \end{pmatrix}, \quad |r^{(a,b)}\succ = \begin{pmatrix} r_1^{(a,b)} \\ r_2^{(a,b)} \\ \vdots \\ r_{N_{(a,b)}}^{(a,b)} \end{pmatrix} \tag{22}$$

with components

$$r_i^{(a,b)} = \varepsilon\, \omega_{(a,b)}\, \delta_{i,1} - \lambda\, V_{i,1}^{(N_{(a,b)},M)}\, \omega_{(a)} - \lambda\, V_{i,1}^{(N_{(a,b)},N)}\, \omega_{(b)} = r_i^{(a,b)}(\varepsilon, \vec{\omega})\,. \tag{23}$$

Treating, temporarily, the two not yet specified quantities $\Omega_{(a)}$ and $\Omega_{(b)}$ as adjustable matrix-regularization parameters, and replacing the $L = 1$ auxiliary matrix (A9) by its partitioned $L = 2$ counterpart

$$\mathcal{A}(\varepsilon) = \begin{bmatrix} A(M,\varepsilon) & 0 \\ 0 & A(N,\varepsilon) \end{bmatrix}$$

we are just left with the problem of finding a suitable $L = 2$ analogue of relation (A11).

A key to the resolution of the puzzle is found in Equation (21) and in its partitioned direct-sum extension

$$\Pi^{(M \oplus N)} = \begin{bmatrix} \Pi^{(a)} & 0 \\ 0 & \Pi^{(b)} \end{bmatrix}\,.$$

Using this symbol, we can now rewrite our homogeneous Schrödinger Equation (15) in the inhomogeneous matrix-inversion representation

$$\left(\mathcal{A}^{-1}(\varepsilon) + \lambda\, V^{(K)}\, \Pi^{(M \oplus N)} \right) \begin{pmatrix} |\vec{y}^{(a)}\succ \\ |\vec{y}^{(b)}\succ \end{pmatrix} = \begin{pmatrix} |r^{(a)}\succ \\ |r^{(b)}\succ \end{pmatrix} \tag{24}$$

or, equivalently,

$$\left(I + \lambda\, \mathcal{A}(\varepsilon)\, V^{(K)}\, \Pi^{(M \oplus N)} \right) |\vec{y}\rangle = \mathcal{A}\left(\varepsilon\right)|r\rangle. \tag{25}$$

Once we drop the redundant superscripts, and once we add the relevant parameter-dependences in (25) we obtain relation

$$(I + \lambda\, \mathcal{A}(\varepsilon)\, V\, \Pi)\, |\vec{y}\rangle = \mathcal{A}(\varepsilon)\, |r(\lambda,\varepsilon,\vec{\omega})\rangle. \tag{26}$$

This is our ultimate, iteration-friendly exact form of our perturbed Schrödinger equation.

4.3. Solutions

Equation (26) yields the ket-vector part of the solution in closed form,

$$|\vec{y}\rangle = (I + \lambda\, \mathcal{A}(\varepsilon)\, V\, \Pi)^{-1}\, \mathcal{A}(\varepsilon)\, |r(\lambda,\varepsilon,\vec{\omega})\rangle = |\vec{y}^{(solution)}(\lambda,\varepsilon,\vec{\omega})\rangle. \tag{27}$$

In the small-perturbation regime the latter formula may be given the conventional Taylor-series form with

$$|\vec{y}^{(solution)}(\lambda,\varepsilon,\vec{\omega})\rangle = \mathcal{A}(\varepsilon)\, |r(\lambda,\varepsilon,\vec{\omega})\rangle - \lambda\, \mathcal{A}(\varepsilon)\, V\, \Pi\, \mathcal{A}(\varepsilon)\, |r(\lambda,\varepsilon,\vec{\omega})\rangle +$$

$$+ \lambda^2\, \mathcal{A}(\varepsilon)\, V\, \Pi\, \mathcal{A}(\varepsilon)\, V\, \Pi\, \mathcal{A}(\varepsilon)\, |r(\lambda,\varepsilon,\vec{\omega})\rangle - \ldots. \tag{28}$$

Naturally, the construction is not yet finished because what is missing is the guarantee of equivalence between the eigenvalue problem (15) and its matrix-inversion reformulation (24) containing two redundant parameters. This is the last obstacle, easily circumvented by our setting, in solution (28), both of the redundant upper-case constants $\Omega_{(a)} = y_M^{(a)}$ and $\Omega_{(b)} = y_N^{(b)}$ equal to zero. In the light of explicit formula (28), this requirement is equivalent to the pair of relations

$$y_M^{(solution)}(\lambda,\varepsilon,\omega_{(a)},\omega_{(b)}) = 0, \quad y_{M+N}^{(solution)}(\lambda,\varepsilon,\omega_{(a)},\omega_{(b)}) = 0. \tag{29}$$

Both left-hand-side functions of the three unknown quantities ε, $\omega_{(a)}$ and $\omega_{(b)}$ are available, due to formula (28), in closed form. Although both can vary with the three independent unknowns (i.e., with $\varepsilon, \omega_{(a)}$ and $\omega_{(b)}$), one of these variables merely plays the role of an optional normalization constant so that we may set, say, $\omega_{(a)}^2 + \omega_{(b)}^2 = 1$. Thus, the implicit version of the perturbation-expansion construction of bound states is completed.

5. Schrödinger Equation in Leading-Order Approximation

Two coupled equations (29) determine the bound state. In the spirit of perturbation theory one may expect that the perturbations happen to be, in some sense, small. At the same time, even the analysis of the comparatively elementary $L = 1$ secular Equation (A16) determining the single free variable (viz., the energy) led to the necessity of a strongly counterintuitive scaling (A18) reflecting, near the extreme EP boundary of unitarity, the strong anisotropy of the geometry of the physical Hilbert space. Naturally, at least comparable complications must be expected to be encountered during the analysis of the more complicated set of two coupled equations (29) representing the secular equation in its exact $L = 2$ version, constructed as particularly suitable for systematic approximations.

5.1. Generic Case: Perturbations without Vanishing Elements

Even the most drastic truncation of the formal power series (28) yields already a nontrivial ket vector

$$|\vec{y}^{(solution)}(\lambda,\varepsilon,\vec{\omega})\rangle = \mathcal{A}(\varepsilon)\, |r(\lambda,\varepsilon,\vec{\omega})\rangle. \tag{30}$$

Needless to add that what must vanish are the auxiliary variables $\Omega_{(a,b)}$ *alias* two functions which are available in closed form. Thus, we must solve the following two simplified equations

$$\sum_{k=1}^{N_{(\varrho)}} A_{N_{(\varrho)},k}^{(\varrho)}(\varepsilon)\, r_k^{(\varrho)}(\lambda,\varepsilon,\omega_{(a)},\omega_{(b)}) = 0, \qquad \varrho = a,b. \tag{31}$$

After the insertion of the respective matrix elements $A_{N_{(a,b)},k}^{(a,b)}(\varepsilon)$ [cf. Equation (A9)] we obtain the pair of relations

$$\left[\sum_{m=0}^{M-1} \epsilon^m\, \lambda\, V_{M-m,1}^{(M,M)}\right] \omega_{(a)} + \left[\sum_{m=0}^{M-1} \epsilon^m\, \lambda\, V_{M-m,1}^{(M,N)}\right] \omega_{(b)} = \epsilon^M\, \omega_{(a)}, \tag{32}$$

$$\left[\sum_{n=0}^{N-1} \epsilon^n\, \lambda\, V_{N-n,1}^{(N,M)}\right] \omega_{(a)} + \left[\sum_{n=0}^{N-1} \epsilon^n\, \lambda\, V_{N-n,1}^{(N,N)}\right] \omega_{(b)} = \epsilon^N\, \omega_{(b)}. \tag{33}$$

Obviously, this set can be read as a generalized eigenvalue problem which determines generalized eigenvectors $\vec{\omega}$ at a K-plet of eigenenergies ϵ which are all defined as roots of the corresponding generalized secular determinant.

5.2. Hierarchy of Relevance and Reduced Approximations

In the generic case one must assume that the matrix elements of the perturbation do not vanish and that λ is small, i.e., that also the eigenvalues ε remain small. This enables one to omit all the asymptotically subdominant corrections and to consider just the linear algebraic system

$$\left[\lambda\, V_{M,1}^{(M,M)} - \epsilon^M\right] \omega_{(a)} + \lambda\, V_{M,1}^{(M,N)}\, \omega_{(b)} = 0, \tag{34}$$

$$\lambda\, V_{N,1}^{(N,M)}\, \omega_{(a)} + \left[\lambda\, V_{N,1}^{(N,N)} - \epsilon^N\right] \omega_{(b)} = 0. \tag{35}$$

The solution of these two simplified coupled linear relations exists if and only if the determinant of the system vanishes,

$$\det \begin{bmatrix} \lambda\, V_{M,1}^{(M,M)} - \epsilon^M & \lambda\, V_{M,1}^{(M,N)} \\ \lambda\, V_{N,1}^{(N,M)} & \lambda\, V_{N,1}^{(NN)} - \epsilon^N \end{bmatrix} = 0. \tag{36}$$

Thus, an ordinary eigenvalue problem is encountered when $M = N$.

Lemma 1. *In the generic equipartitioned cases with $M = N \geq 3$ the spectrum ceases to be all real under bounded perturbations. The loss of unitarity is encountered.*

Proof. Both roots $\epsilon^N = \lambda\, x$ of the exactly solvable quadratic algebraic secular Equation (36) may be guaranteed to be real in a certain domain of parameters. Still, some of the energies themselves are necessarily complex since $\epsilon = \sqrt[N]{\lambda\, x}$ is an N-valued function with values lying on a complex circle. $\square$

In the other, non-equipartitioned dynamical scenarios with, say, $M > N$, the behavior of the system in an immediate vicinity of its EP extreme is still determined by the asymptotically dominant part of the secular equation. After an appropriate modification, the above proof still applies.

Lemma 2. *In the generic case with $M > N \geq 3$ we get, from the dominant part of the generalized eigenvalue problem (36), a subset (N-plet) of asymptotically dominant eigenvalues $\varepsilon = \mathcal{O}(\lambda^{1/N})$ which cannot be all real.*

5.3. Unitary Case: Re-Scaled Perturbations

The loss of unitarity occurring in the $L = 1$ models was associated with the use of the too broad a class of norm-bounded perturbations. From our preliminary results described in preceding subsection one can conclude that a similar loss of unitarity may be also expected to occur, in the generic case, at $L = 2$. Indeed, the anisotropy of the physical Hilbert space which reflects the influence of an

EP-related singularity of the Hamiltonian may be expected to lead again to a selective enhancement of the weight of certain specific matrix elements of perturbations $\lambda\, V^{(K)}$.

Once we recall the $L = 1$ scenario of Appendix B.3 we immediately imagine that the main source of the apparent universality of the instability under norm-bounded perturbations should be sought, paradoxically, in the routine but, in our case, entirely inadequate norm-boundedness assumption itself. Indeed, in a way documented by Lemmas 1 and 2, the loss of the reality of spectra may directly be attributed to the conventional and comfortable but entirely random, unfounded and formal assumption of the uniform boundedness of the matrix elements of the perturbations.

Most easily the latter result may be illustrated using the drastically simplified version (36) of the leading-order secular equation. Dominant role is played there by the quadruplet of matrix elements $V_{(P,1)}^{(P,Q)}$ with superscripts P and Q equal to M or N. Once they are assumed λ-independent and non-vanishing, the leading-order energies read $\epsilon = \sqrt[N]{\lambda\, x}$. Thus, at any $N \geq 3$ their N-plet forms an equilateral N-angle in the complex plane of λ.

The latter observation inspires a remedy. In a way eliminating the $N \geq 3$ complex-circle obstruction one simply has to re-scale the energies as well as all the relevant matrix elements of the perturbation. In this manner the ansatz

$$\epsilon = \epsilon(E) = \sqrt{\lambda}\, E \tag{37}$$

opens the possibility of the spectrum being real. Another multiplet of postulates

$$V_{M-m,1}^{(M,M)} = \lambda^{(M-m)/2}\, W_{M-m,1}^{(M,M)}, \quad V_{M-m,1}^{(M,N)} = \lambda^{(M-m)/2}\, W_{M-m,1}^{(M,N)}, \quad m = 0,1,\ldots,M-1, \tag{38}$$

and

$$V_{N-n,1}^{(N,M)} = \lambda^{(N-n)/2}\, W_{N-n,1}^{(N,M)}, \quad V_{N-n,1}^{(N,N)} = \lambda^{(N-n)/2}\, W_{N-n,1}^{(N,N)}, \quad n = 0,1,\ldots,N-1 \tag{39}$$

now contains a new partitioned matrix $W = W^{(K)}$ which is assumed uniformly bounded.

Lemma 3. *There always exists a non-empty $(2M + 2N)$-dimensional domain $\mathcal{D}$ of the "physical" matrix elements of W for which the leading-order spectrum is all real and non-degenerate, i.e., in the language of physics, tractable as stable bound-state energies.*

Proof. The main consequence of the amended, necessary-condition perturbation-smallness requirements (38) and (39) is that in our initial, unreduced leading-order generalized eigenvalue problem (32) + (33), all terms in the sums become of the same order of magnitude. By the scaling we managed to eliminate the explicit presence of the measure of smallness λ. In other words the input information about dynamics is formed now by the re-scaled perturbation matrix W which offers an $(2M + 2N)$-plet of free $\mathcal{O}(1)$ parameters. At the same time, the spectral-definition output is given by the $M + N$ roots of the corresponding secular equation, i.e., by the roots of an $(M + N)$-th-degree polynomial in E, with all its separate coefficients bounded and, in general, non-vanishing. Under these conditions the assertion of the lemma is obvious. $\square$

6. Discussion

6.1. Schrödinger Picture and Quasi-Hermitian Hamiltonians

In the context of the new, Bender- and Boettcher-inspired paradigm the non-Hermiticity of $H(\gamma)$ in $\mathcal{K}$ may be considered compatible with unitarity whenever the spectrum itself is found real. The explanation of the apparent paradox is easy: Under certain reasonable mathematical assumptions (cf. [21]) one can find a non-unitary invertible mapping $\Omega = \Omega(\gamma)$ with property $\Omega^\dagger \Omega = \Theta \neq I$ which makes our Hamiltonian self-adjoint,

$$H(\gamma) \to \mathfrak{h}(\gamma) = \Omega(\gamma)\, H(\gamma)\, \Omega(\gamma) = \mathfrak{h}^\dagger(\gamma)\,. \tag{40}$$

In practice the idea enables one to avoid the use of the "conventional" Hamiltonian $\mathfrak{h}(\gamma)$ (self-adjoint, by construction, in another Hilbert space $\mathcal{L}$) whenever it happens to be "prohibitively complicated". In the literature such a type of simplification of calculations is usually attributed to Dyson [22]. Expectedly, the strategy (also known as preconditioning) proved efficient as a tool of construction of bound states in nuclear physics [21]. In spite of a certain initial doubts [23,24], the approach also proved applicable in the quantum physics of scattering [25,26].

From a more abstract theoretical point of view the reference to the "Dyson's" isospectral mapping (40) becomes redundant when we reclassify the space $\mathcal{K}$ as "unphysical", and when we redefine its inner product yielding another, "physical" Hilbert space $\mathcal{H}$,

$$\langle \psi_1 | \psi_2 \rangle_{\mathcal{H}} = \langle \psi_1 | \Theta | \psi_2 \rangle_{\mathcal{K}}\,, \quad |\psi_{1,2}\rangle \in \mathcal{K}\,. \tag{41}$$

A new, equivalent, "two-Hilbert-space" version of the Schrödinger picture is obtained in which the physics described in the correct Hilbert space $\mathcal{H}$ is "translated" to its mathematically easier representation in $\mathcal{K}$. In this sense, the self-adjointness of hypothetical $\mathfrak{h}$ in hypothetical $\mathcal{L}$ is found equivalent to the self-adjointness of H in $\mathcal{H}$, represented by the relation

$$H^\dagger(\gamma)\, \Theta(\gamma) = \Theta(\gamma)\, H(\gamma)\,. \tag{42}$$

Dieudonné [27] and Scholtz et al [21] suggested to call relation (42) the quasi-Hermiticity of H in $\mathcal{K}$. In the context of quantum phenomenology a decisive amendment of the formalism may be seen in the split of the description of dynamics with information carried by the two operators $H(\gamma)$ and $\Theta(\gamma)$ defined in $\mathcal{K}$ in place of one (viz., of $\mathfrak{h}(\gamma)$ living in $\mathcal{L}$).

6.2. Non-Hermitian Degeneracies with $L > 2$

After the present decisive clarification of the possibility of the replacement of the $L = 1$ formalism by its $L = 2$ generalization, the next move to the further, $L > 2$ dynamical scenarios is now an almost elementary exercise. Indeed, in Sections 3–5 it would be sufficient to move from the $L = 2$ partitioning of $K = N_{(a)} + N_{(b)}$ to its arbitrary $L > 2$ analogues (A1) and to the related partitioned vector sets [sampled, e.g., by Equation (7)] and matrices [sampled, e.g., by Equation (8)]. The $L = 2$ doublets of the eligible superscripts $^{(a)}$ and $^{(b)}$ marking the curly kets [cf. (17) or (31), etc.] may be very easily extended to the L-plets with $L > 2$, etc.

Along these lines the form of our basic power-series expansion of the perturbed bound-state kets (28) remains unchanged. The related $L = 2$ compatibility constraint (29) must only be replaced by an L-plet of equations

$$y^{(solution)}_{N_{(a_j)}}(\lambda, \varepsilon, \vec{\omega}) = 0\,, \quad j = 1, 2, \ldots, L \tag{43}$$

where the unknown vector $\vec{\omega}$ has L components.

6.3. The Next-to-Leading-Order Approximation

In our present paper we did not pay too much attention to the next-to-leading-order (NLO) approximation where one would have to set

$$|\vec{y}^{(solution)}(\lambda, \varepsilon, \vec{\omega})\rangle = \mathcal{A}(\varepsilon)\, |r(\lambda, \varepsilon, \vec{\omega})\rangle - \lambda\, \mathcal{A}(\varepsilon)\, V\, \Pi\, \mathcal{A}(\varepsilon)\, |r(\lambda, \varepsilon, \vec{\omega})\rangle\,. \tag{44}$$

Our main reason was that such a generalization would make the associated compatibility conditions (43) perceivably more complicated. Indeed, in contrast to the leading-order case, an almost inessential improvement of the insight in the qualitative features of the quantum system in question would be accompanied, in the NLO formulae, by the emergence of multiple new matrix elements of perturbations $\lambda V^{(K)}$ including even the terms which would be quadratic functions of these matrix elements.

This being said, it is necessary to add that even on the pragmatic and qualitative level, the omission of the NLO corrections only seems completely harmful in the open-system setting using bounded matrices $\lambda V^{(K)}$ where one does not insist on having the strictly real spectrum. The point is that whenever one must guarantee the unitarity of the system, the class of the admissible perturbations must be further restricted. In this sense, unfortunately, an analysis using NLO might prove necessary. Indeed, the use and precision of the leading-order approximation need not be sufficiently reliable in general. Even a quick glimpse at the underlying assumptions (38) and (39) reveals that the leading-order approximation does not incorporate the influence of a large subset of the matrix elements of $\lambda V^{(K)}$. At the same time, one must keep in mind that in our present approach the dimension of the matrices K has been assumed finite. For this reason, in the case of doubts, a turn to the more universal and brute-force numerical methods might prove to be, in practical calculations, a reasonable alternative to the rather lengthy and complicated NLO calculations.

Funding: This work is supported by the Nuclear Physics Institute in Řež and by the Faculty of Science of the University of Hradec Králové.

Acknowledgments: The author acknowledges the financial support from the Excellence project PřF UHK 2020 Nr. 2212.

Conflicts of Interest: The author declares no conflict of interest.

Appendix A. An Exhaustive Classification of the Degeneracies of Exceptional Points

Up to an arbitrary permutation, every partitioning

$$K = M_1 + M_2 + \ldots + M_L \tag{A1}$$

of the full matrix dimension characterizes a different system and its NHD limit. Thus, we must postulate, say,

$$M_1 \geq M_2 \geq \ldots \geq M_L \geq 2$$

and introduce the schematic multi-indices $\boxed{M_1 + M_2 + \ldots + M_L}$ in a way illustrated in Table A1.

Table A1. Eligible EP-related partitioning.

K	List
2	2
3	3
4	4 | 2 + 2
5	5 | 3 + 2
6	6 | 4 + 2 | 3 + 3 | 2 + 2 + 2
7	7 | 5 + 2 | 4 + 3 | 3 + 2 + 2
8	8 | 6 + 2 | 5 + 3 | 4 + 4 | 4 + 2 + 2 | 3 + 3 + 2 | 2 + 2 + 2 + 2
9	9 | 7 + 2 | 6 + 3 | 5 + 4 | 5 + 2 + 2 | 4 + 3 + 2 | 3 + 3 + 3 | 3 + 2 + 2 + 2
10	10 | 8 + 2 | 7 + 3 | 6 + 4 | 6 + 2 + 2 | 5 + 5 | 5 + 3 + 2
	4 + 4 + 2 | 4 + 3 + 3 | 4 + 2 + 2 + 2 | 3 + 3 + 2 + 2 | 2 + 2 + 2 + 2 + 2
$\vdots$	$\ldots$

The Table indicates that the nontrivial $L > 1$ partitioning not containing trivial items $M_j = 1$ only exists at $K \geq 4$. The value of the count $\mathcal{N}(K)$ of the separate nontrivial, $L \geq 2$ items is seen to exceed one only at $K = 6$. Nevertheless, this count starts growing quickly at the larger Ks (see Table A2). Thus, the current practice of studying just the EPK models with $L = 1$ misses in fact the huge majority of the alternative NHD scenarios. Marginally, it is worth adding that the counts $\mathcal{N}(K)$ form a well-known sequence. In the open-access on-line encyclopedia of integer sequences [28] it is assigned the code number A083751.

Table A2. The sequence of counts $\mathcal{N}(K)$ with $K = 2, 3, \ldots$.

$$0, 0, 1, 1, 3, 3, 6, 7, 11, 13, 20, 23, 33, 40, 54, 65, 87, 104, 136, 164,$$
$$209, 252, 319, 382, 477, 573, 707, 846, 1038, 1237, 1506, 1793, \ldots .$$

For our present purposes, the asymptotic growth of sequence $\mathcal{N}(K)$ (which is slightly slower than exponential) as well as its precise mathematical definition is less relevant. At the realistic, not too large dimensions K, for example, we might also use some alternative definitions. One of them leads to values $\mathcal{N}(K)$ equal to the first differences (diminished by one) of the special partitions (of the code number A000041, cf. [29]), or to the first differences of the numbers of trees of diameter four (see the integer sequence with code number A000094 in [30]). Nevertheless, irrespectively of the choice of definition let us point out that in the NHD vicinity, the partitioning multi-indices will classify the phenomenologically non-equivalent physical systems in general. As long as the superscripts $^{(K)}$ are in fact redundant, we will omit or replace them by the more relevant information about the partitioning, therefore. In particular, symbol

$$\mathcal{J}^{(M_1 \oplus M_2 \oplus \ldots \oplus M_L)}(\eta) \tag{A2}$$

will represent the general block-diagonal canonical representation $\mathcal{J}^{(K)}(\eta)$ of the Hamiltonian defined in Equation (8).

Appendix B. Perturbation Theory Near Non-Degenerate Exceptional Points

Appendix B.1. The Choice of Basis at $L = 1$

Relation (4) can be read as a definition of transition matrix responsible for the canonical K by K Jordan-block representation

$$J^{(K)}(\eta) = \left[Q^{(EPK)} \right]^{-1} H(\gamma^{(EPK)}) \, Q^{(EPK)} \tag{A3}$$

of the EPK $L = 1$ limit of any given non-Hermitian but $\mathcal{PT}$-symmetric Hamiltonian of phenomenological relevance. From this point of view one can extend the same transformation (i.e., for matrices with $K < \infty$, a mere choice of the basis in Hilbert space) to a vicinity of the EPK singularity. This yields the Jordan block matrix plus perturbation,

$$\left[Q^{(EPK)} \right]^{-1} H^{(K)}(\gamma) \, Q^{(EPK)} = J^{(K)}(\eta) + \lambda \, V^{(K)}(\gamma). \tag{A4}$$

Such a definition contains a redundant but convenient measure $\lambda = \mathcal{O}(\gamma - \gamma^{(EPK)})$ of the smallness of perturbation.

Due to the conventional postulate of having a specific one-parametric family of Hamiltonians $H^{(K)}(\gamma)$ given in advance, the introduction of the concept of perturbation in (A4) is just a formal step. Nevertheless, the interaction term itself could be also reinterpreted as a model-independent random perturbation which carries the input dynamical information. From such a perspective every preselected perturbation term defines a different Hamiltonian $\widetilde{H^{(K)}})$ which merely coincides with

$H^{(K)}(\gamma)$ in the NHD limit $\gamma \to \gamma^{(EP)}$ (a few exactly solvable samples of such a truly remarkable Hamiltonian matching may be found in [15]). This means that via relation

$$J^{(K)}(\eta) + \lambda V^{(K)} = \left[Q^{(EPK)}\right]^{-1} \widetilde{H^{(K)}}) \, Q^{(EPK)} \tag{A5}$$

[i.e., via a mere reordering of relation (A4)] we obtain a new picture of physics in which the resulting tilded Hamiltonian with real spectrum need not be $\mathcal{PT}$-symmetric at all.

Appendix B.2. The Description of the Unfolding of the Degeneracy at $L = 1$

A specific constructive extension of the latter observation has been presented in our recent paper [18]. We were able there to prove that in the NHD dynamical regime, the mere boundedness of the norm of matrix $V^{(K)}$ together with the smallness of parameter λ still *do not* guarantee the survival of the unitarity of the system after perturbation. We showed there (cf. also [14]) that one can guarantee the absence of a "quantum catastrophe" (i.e., of an abrupt change of some of the system's observable features, see [31–33]) only via a certain self-consistent revision of the criteria of smallness of matrix $V^{(K)}$.

Having in mind the parameter-independence and invertibility of the transformation matrix $Q^{(EPK)}$ the quantification of the influence of the perturbation is of enormous interest, among others, in the analysis of stability of the system in question [12]. This was the reason we also addressed the $L = 1$ perturbative bound-state problem

$$\left[J^{(K)}(0) + \lambda V^{(K)}\right] |\Psi\rangle = \epsilon |\Psi\rangle, \tag{A6}$$

with $\eta = 0$ in Ref. [17]. With the ket-vector subscripts j in $|\Psi_j\rangle$ running from 1 to K we fixed the norm (by setting $|\Psi_1\rangle = 1$), and we relocated the first column of Equation (A6), viz., vector

$$\vec{r} = \vec{r}(\lambda) = \begin{pmatrix} \epsilon - \lambda V_{1,1} \\ -\lambda V_{2,1} \\ \vdots \\ -\lambda V_{K,1} \end{pmatrix} \tag{A7}$$

to the right-hand side of the equation. Then we restored the comfortable square-matrix form of the equation via its two further equivalent modifications. First we added a new, temporarily undetermined auxiliary component Ω_K to an "upgrade" of the wave function

$$\vec{y} = \begin{pmatrix} y_1 \\ y_2 \\ \vdots \\ y_{K-1} \\ y_K \end{pmatrix} = \begin{pmatrix} |\Psi_2\rangle \\ |\Psi_3\rangle \\ \vdots \\ |\Psi_K\rangle \\ \Omega_K \end{pmatrix}. \tag{A8}$$

Subsequently, an introduction of the following auxiliary lower-triangular K by K matrix

$$A = A(K,\epsilon) = \begin{pmatrix} 1 & 0 & 0 & \dots & 0 \\ \epsilon & 1 & 0 & \ddots & \vdots \\ \epsilon^2 & \epsilon & \ddots & \ddots & 0 \\ \vdots & \ddots & \ddots & 1 & 0 \\ \epsilon^{K-1} & \dots & \epsilon^2 & \epsilon & 1 \end{pmatrix} \tag{A9}$$

and of its two-diagonal inverse

$$
A^{-1} = \begin{pmatrix}
1 & 0 & 0 & \cdots & 0 \\
-\epsilon & 1 & 0 & \ddots & \vdots \\
0 & -\epsilon & \ddots & \ddots & 0 \\
\vdots & \ddots & \ddots & 1 & 0 \\
0 & \cdots & 0 & -\epsilon & 1
\end{pmatrix}
\tag{A10}
$$

accompanied by a parallel formal upgrade of the interaction matrix,

$$
V^{(K)} \; \rightarrow \; Z = \begin{pmatrix}
V_{1,2} & V_{1,3} & \cdots & V_{1,K} & 0 \\
V_{2,2} & V_{2,3} & \cdots & V_{2,K} & 0 \\
\cdots & \cdots & \cdots & \vdots & \vdots \\
V_{K,2} & V_{K,3} & \cdots & V_{K,K} & 0
\end{pmatrix}
\tag{A11}
$$

enabled us to rewrite our initial homogeneous Schrödinger Equation (A6), in the last step of the construction of its solution, in an equivalent matrix-inversion form

$$
(A^{-1} + \lambda Z)\vec{y} = \vec{r}
\tag{A12}
$$

or, better,

$$
(I + \lambda A Z)\vec{y} = A\vec{r}
\tag{A13}
$$

accompanied by the innocent-looking but important self-consistence constraint

$$
\Omega_K = 0.
\tag{A14}
$$

In the leading-order application of the recipe we then returned to the slightly vague assumption of the "sufficient smallness" of the perturbation. On these grounds we recalled the formal Taylor-series expansion of the resolvent which yielded the closed formula

$$
\vec{y}^{(solution)}(\epsilon) = A\vec{r} - \lambda A Z A\vec{r} + \lambda^2 A Z A Z A\vec{r} - \dots
\tag{A15}
$$

for the modified wave function. It contained a free parameter ϵ which had to be fixed via the supplementary secular Equation (A14). In the light of the Taylor-series formula (A15), such a secular equation now acquires the K-th-vector-component form

$$
y_K^{(solution)}(\epsilon) = 0.
\tag{A16}
$$

of an explicit transcendental equation for the energies ϵ.

Appendix B.3. Unitary-Evolution Process of Unfolding at $L = 1$

The latter constraint (A14) plays the role of an implicit definition of the spectrum. The K-plet of roots $\epsilon_n = \epsilon_n(\lambda)$, $n = 1, 2, \dots, K$ represents the bound-state energies. After the truncation of the series, just approximate solutions are being obtained. In Ref. [17], incidentally, even the leading-order roots were found complex in general.

This observation was interpreted as indicating that in an immediate EPK vicinity the norm-bounded perturbations λV should still be considered, in the unitary theory, "inadmissibly large". The non-unitary, open-quantum system interpretation of the perturbations proved needed forcing the system to perform, at an arbitrarily small but non-vanishing $\lambda \neq 0$, an abrupt quantum phase transition.

Incidentally, qualitatively the same conclusions were also obtained in the above-mentioned more concrete study [11] of the specific Bose–Hubbard model in its EPK dynamical regime. The resolution of an apparent universal-instability paradox was provided in our subsequent study [18] in which we studied the underlying exact as well as approximate secular equations in more detail. Our ultimate conclusion was that the necessary smallness condition specifying the class of the admissible, unitarity non-violating perturbations does not involve their upper-triangular matrix part at all. In contrast, their lower-triangular matrix part must be given the following, matrix-element-dependent form

$$
\lambda\,V^{(K)}_{(admissible)} = \begin{bmatrix}
\lambda^{1/2}\mu_{11} & 0 & \cdots & 0 & 0 & 0 \\
\lambda\,\mu_{21} & \lambda^{1/2}\mu_{22} & \cdots & 0 & 0 & 0 \\
\lambda^{3/2}\mu_{31} & \lambda\,\mu_{32} & \ddots & \vdots & \vdots & 0 \\
\lambda^{2}\mu_{41} & \lambda^{3/2}\mu_{42} & \ddots & \ddots & 0 & 0 \\
\vdots & \vdots & \ddots & \lambda\,\mu_{K-1K-2} & \lambda^{1/2}\mu_{K-1K-1} & 0 \\
\lambda^{K/2}\mu_{K1} & \lambda^{(K-1)/2}\mu_{K2} & \cdots & \lambda^{3/2}\mu_{KK-2} & \lambda\,\mu_{KK-1} & \lambda^{1/2}\mu_{KK}
\end{bmatrix}. \tag{A17}
$$

During the decrease of $\lambda \to 0$, all the variable lower-triangle matrix-element parameters must remain bounded, $\mu_{j,k} = \mathcal{O}(1)$. In other words, as long as we are working in a specific, fixed "unperturbed" basis, the matrix structure (A17) may be interpreted as manifesting a characteristic anisotropy and the hierarchically ordered weights of influence of the separate matrix elements. Indeed, we may re-scale

$$
\lambda\,V^{(K)}_{(admissible)} = \lambda^{1/2}\,B(\lambda)\,V^{(reduced)}\,B^{-1}(\lambda) \tag{A18}
$$

where $B(\lambda)$ would be a diagonal matrix with elements $B_{jj}(\lambda) = \lambda^{j/2}$ and where the reduced perturbation matrix would be bounded, $V^{(reduced)}_{jk} = \mathcal{O}(1)$.

On this necessary-condition background valid at all dimensions K, the samples of sufficient conditions retain a purely numerical trial-and-error character, with the small$-K$ non-numerical exceptions discussed, in [18], for the matrix dimensions up to $K = 5$.

References

1. Bender, C.M., Boettcher, S. Real spectra in non-Hermitian Hamiltonians having $\mathcal{PT}$ symmetry. *Phys. Rev. Lett.* **1998**, *80*, 5243.
2. Bender, C.M. Making Sense of Non-Hermitian Hamiltonians. *Rep. Prog. Phys.* **2007**, *70*, 947–1018.
3. Mostafazadeh, A. Pseudo-Hermitian Quantum Mechanics. *Int. J. Geom. Meth. Mod. Phys.* **2010**, *7*, 1191–1306.
4. Bagarello, F.; Gazeau, J.-P.; Szafraniec, F.; Znojil, M. (Eds.) *Non-Selfadjoint Operators in Quantum Physics: Mathematical Aspects*; Wiley: Hoboken, NJ, USA, 2015.
5. Bender, C.M. (Ed.) *PT Symmetry in Quantum and Classical Physics*; World Scientific: Singapore, 2018.
6. Berry, M.V. Physics of non-Hermitian degeneracies. *Czech. J. Phys.* **2004**, *54*, 1039–1047.
7. Kato, T. *Perturbation Theory for Linear Operators*; Springer: Berlin, Germany, 1966.
8. Bloch, I.; Hänsch, T.W.; Esslinger, T. Atom laser with a cw output coupler. *Phys. Rev. Lett.* **1999**, *82*, 3008.
9. Hiller, M.; Kottos, T.; Ossipov, A. Bifurcations in resonance widths of an open Bose-Hubbard dimer. *Phys. Rev. A* **2006**, *73*, 063625.
10. Teimourpour, M.H.; Zhong, Q.; Khajavikhan, M.; El-Ganainy, R. Higher Order EPs in iscrete Photonic Platforms. In *Parity-Time Symmetry and Its Applications*; Christodoulides, D., Yang, J.-K., Eds.; Springer: Singapore, 2018; pp. 261–276.
11. Graefe, E.M.; Günther, U.; Korsch, H.J.; Niederle, A.E. A non-Hermitian PT symmetric Bose-Hubbard model: Eigenvalue rings from unfolding higherorder exceptional points. *J. Phys. A Math. Theor.* **2008**, *41*, 255206.
12. Krejčiřík, D.; Siegl, P.; Tater, M.; Viola, J. Pseudospectra in non-Hermitian quantum mechanics. *J. Math. Phys.* **2015**, *56*, 103513.

13. Moiseyev, N. *Non-Hermitian Quantum Mechanics*; CUP: Cambridge, UK, 2011.
14. Znojil, M.; Růžička, F. Nonlinearity of perturbations in PT-symmetric quantum mechanics. *J. Phys. Conf. Ser.* **2019**, *1194*, 012120.
15. Znojil, M. Passage through exceptional point: Case study. *Proc. R. Soc. A Math. Phys. Eng. Sci.* **2020**, *476*, 20190831.
16. Znojil, M.; Borisov, D.I. Anomalous mechanisms of the loss of observability in non-Hermitian quantum models. *Nucl. Physci. B* **2020**, *957*, 115064.
17. Znojil, M. Admissible perturbations and false instabilities in PT-symmetric quantum systems. *Phys. Rev. A* **2018**, *97*, 032114.
18. Znojil, M. Unitarity corridors to exceptional points. *Phys. Rev. A.* **2019**, *100*, 032124.
19. Seyranian, A.P.; Mailybaev, A.A. *Multiparameter Stability Theory with Mechanical Applications*; World Scientific: Singapore, 2003.
20. Christodoulides, D.; Yang, J.-K. (Eds.) *Parity-Time Symmetry and Its Applications*; Springer: Singapore, 2018.
21. Scholtz, F.G.; Geyer, H.B.; Hahne, F.J.W. Quasi-Hermitian Operators in Quantum Mechanics and the Variational Principle. *Ann. Phys.* **1992**, *213*, 74.
22. Dyson, F.J. General theory of spin-wave interactions. *Phys. Rev.* **1956**, *102*, 1217–1230
23. Jones, H.F. Scattering from localized non-Hermitian potentials. *Phys. Rev. D* **2007**, *76*, 125003.
24. Jones, H.F. Interface between Hermitian and non-Hermitian Hamiltonians in a model calculation. *Phys. Rev. D* **2008**, *78*, 065032.
25. Znojil, M. Discrete PT-symmetric models of scattering. *J. Phys. A Math. Theor.* **2008**, *41*, 292002.
26. Znojil, M. Scattering theory using smeared non-Hermitian potentials. *Phys. Rev. D* **2009**, *80*, 045009.
27. Dieudonné, J. Quasi-Hermitian operators. In Proceedings of the International Symposium on Linear Spaces, Jerusalem, Israel, 5–12 July 1961; Pergamon: Oxford, UK, 1961; pp. 115–122.
28. The On-Line Encyclopedia of Integer Sequences. Available online: http://oeis.org/A083751/ (accessed on 6 November 2019).
29. The On-Line Encyclopedia of Integer Sequences. Available online: http://oeis.org/A000041/ (accessed on 6 November 2019).
30. The On-Line Encyclopedia of Integer Sequences. Available online: http://oeis.org/A000094/ (accessed on 6 November 2019).
31. Znojil, M. Quantum catastrophes: A case study. *J. Phys. A Math. Theor.* **2012**, *45*, 444036.
32. Borisov, D.I. Eigenvalue collision for PT-symmetric waveguide. *Acta Polytech.* **2014**, *54*, 93.
33. Znojil, M.; Borisov, D.I. Two patterns of PT-symmetry breakdown in a non-numerical six-state simulation. *Ann. Phys.* **2018**, *394*, 40.

Article

Deformed Shape Invariant Superpotentials in Quantum Mechanics and Expansions in Powers of $\hbar$

Christiane Quesne

Physique Nucléaire Théorique et Physique Mathématique, Université Libre de Bruxelles,
Campus de la Plaine CP229, Boulevard du Triomphe, B-1050 Brussels, Belgium; cquesne@ulb.ac.be

Received: 30 September 2020; Accepted: 29 October 2020; Published: 10 November 2020

Abstract: We show that the method developed by Gangopadhyaya, Mallow, and their coworkers to deal with (translational) shape invariant potentials in supersymmetric quantum mechanics and consisting in replacing the shape invariance condition, which is a difference-differential equation, which, by an infinite set of partial differential equations, can be generalized to deformed shape invariant potentials in deformed supersymmetric quantum mechanics. The extended method is illustrated by several examples, corresponding both to $\hbar$-independent superpotentials and to a superpotential explicitly depending on $\hbar$.

Keywords: quantum mechanics; supersymmetry; shape invariance; curved space; position-dependent mass

PACS: 03.65.Fd; 03.65.Ge

1. Introduction

Exactly solvable (ES) Schrödinger equations (SE) allow us to understand some physical phenomena and to test some approximation schemes. Supersymmetric quantum mechanics (SUSYQM) [1–4] is known to be a very powerful method for generating such ES models, especially whenever the corresponding potential is (translationally) shape invariant (SI) [5]. SUSYQM may be considered as a modern version of the old Darboux transformation [6] and of the factorization method used by Schrödinger [7–9] and by Infeld and Hull [10].

Some ten years ago, the list of (translational) SI potentials, whose bound-state wavefunctions can be expressed in terms of classical orthogonal polynomials (COP) [1] has been completed (see [11] and references quoted therein) by introducing [12–15] some rational extensions of these potentials, connected with the novel field of exceptional orthogonal polynomials (EOP) [16]. The latter are polynomial sets which are orthogonal and complete, but, in contrast with COP, admit a finite number of gaps in the sequence of their degrees.

ES models for some unconventional SE are also very interesting. These unconventional equations may be of three different kinds. They may occur whenever the standard commutation relations are replaced by deformed ones, associated with nonzero minimal uncertainties in position and/or momentum [17–19], as suggested by several investigations in string theory and quantum gravity [20]. They may also appear whenever the constant mass of the conventional SE is replaced by a position-dependent mass (PDM). The latter is an essential ingredient in the study of electronic properties of semiconductor heterostructures [21,22], quantum wells and quantum dots [23,24], helium clusters [25], graded crystals [26], quantum liquids [27], metal clusters [28], nuclei [29,30], nanowire structures [31], and neutron stars [32]. A third possibility corresponds to the replacement of the Euclidean space by a curved one. The study of the Kepler–Coulomb problem on the sphere dates back to the work of Schrödinger [7], Infeld [33], and Stevenson [34], and was generalized to a hyperbolic space by Infeld and Schild [35]. Since then, many studies have been devoted to this topic (see [36,37]).

As shown elsewhere [38], there are some intimate connections between these three types of unconventional SE, occurring whenever a specific relation exists between the deforming function $f(x)$, the PDM $m(x)$, and the (diagonal) metric tensor $g(x)$. Such unconventional SE may then be discussed in the framework of deformed SUSYQM (DSUSYQM), where the standard SI condition is replaced by a deformed one (DSI) [39,40]. On starting from the known superpotentials of SI potentials [1], a procedure has been devised in [39] to maintain the solvability of the DSI condition, thereby resulting in a list of deformed superpotentials and deforming functions giving rise to bound-state spectra. In such a deformed case, physically acceptable wavefunctions have not only to be square integrable on the defining interval of the potential, but also must ensure the Hermiticity of the Hamiltonian. More recently [41], this list of deformed superpotentials and deforming functions has been completed by considering the case of some rationally-extended potentials, connected with one-indexed families of EOP.

In the case of conventional SUSYQM, Gangopadhyaya, Mallow, and their coworkers proposed an interesting approach to SI potentials, consisting in replacing the SI condition, which is a difference-differential equation, by an infinite set of partial differential equations. The latter is obtained by expanding the superpotential in powers of $\hbar$ and expressing that the coefficient of each power must separately vanish [42]. This procedure enabled them to prove that the SI superpotentials connected with COP are those with no explicit dependence on $\hbar$, while the new ones related to EOP have such an explicit dependence [43]. They also showed that the list of the former given in [1] is complete [44] and constructed a novel example of SI superpotential with an explicit $\hbar$-dependence [45]. Furthermore, they encountered a pathway for going from those superpotentials of [1] corresponding to SE that can be reduced to the confluent hypergeometric equation to those related to SE connected with the hypergeometric equation [46].

It is the purpose of the present paper to propose an extension of the approach of Gangopadhyaya, Mallow, and their coworkers to the case of DSI potentials in DSUSYQM, both without and with explicit dependence of the superpotential on $\hbar$. We plan to illustrate this method by re-examining the known pairs of deformed superpotentials and deforming functions of [39–41] along these lines. In the present work, we restrict ourselves to unbroken DSUSYQM and only consider the discrete part of the spectrum.

After reviewing the general formalism of DSUSYQM and obtaining the DSI condition in Section 2, we will show in Section 3 how to convert such a condition into a set of partial differential equations in the case where the superpotential does not contain any dependence on $\hbar$. The case where the superpotential has such an explicit dependence is then treated in Section 4. Finally, Section 5 contains the conclusion.

2. Deformed Shape Invariance in Deformed Supersymmetric Quantum Mechanics

In DSUSYQM [39–41], a general Hamiltonian H_- is written in terms of linear operators

$$A^{\pm} = A^{\pm}(a) = \mp \hbar \sqrt{f(x)} \frac{d}{dx} \sqrt{f(x)} + W(x,a), \tag{1}$$

where $f(x)$ is some positive-definite function of x, known as the deforming function, and $W(x,a)$ is a real function of x and a parameter a, called the superpotential. Both $f(x)$ and $W(x,a)$ in general depend on some extra parameters. The Hamiltonian H_- is given by

$$H_- = A^+ A^- = -\hbar^2 \sqrt{f(x)} \frac{d}{dx} f(x) \frac{d}{dx} \sqrt{f(x)} + V_-(x,a), \tag{2}$$

where

$$V_-(x,a) = W^2(x,a) - \hbar f(x)\frac{dW(x,a)}{dx}. \tag{3}$$

It may be interpreted [38] as a Hamiltonian describing a PDM system with $m(x) = 1/f^2(x)$, the ordering of the latter and the differential operator d/dx being that proposed by Mustafa and Mazharimousavi [47], or as a Hamiltonian in a curved space with a diagonal metric tensor $g(x) = 1/f^2(x)$.

The product of operators A^-A^+ generates the so-called partner of H_-,

$$H_+ = A^-A^+ = -\hbar^2\sqrt{f(x)}\frac{d}{dx}f(x)\frac{d}{dx}\sqrt{f(x)} + V_+(x,a), \tag{4}$$

with

$$V_+(x,a) = W^2(x,a) + \hbar f(x)\frac{dW(x,a)}{dx}. \tag{5}$$

The pair of Hamiltonians intertwine with A^+ and A^- as

$$A^-H_- = H_+A^-, \qquad A^+H_+ = H_-A^+. \tag{6}$$

The Hamiltonian H_- is assumed to have a ground-state wavefunction $\psi_0^{(-)}(x,a)$, such that

$$A^-\psi_0^{(-)}(x,a) = 0. \tag{7}$$

From (1) and (2), the latter is therefore such that $E_0^{(-)} = 0$ and

$$\psi_0^{(-)}(x,a) = \frac{N_0^{(-)}}{\sqrt{f(x)}}\exp\left(-\int^x \frac{W(x',a)}{\hbar f(x')}dx'\right), \tag{8}$$

where $N_0^{(-)}$ is the normalization coefficient.

The intertwining relations (6) then imply the following isospectrality relationships among the eigenvalues and eigenfunctions of the two partners,

$$E_{n+1}^{(-)} = E_n^{(+)}, \tag{9}$$

$$\psi_n^{(+)}(x,a) = \frac{A^-}{\sqrt{E_n^{(+)}}}\psi_{n+1}^{(-)}(x,a), \qquad \psi_{n+1}^{(-)}(x,a) = \frac{A^+}{\sqrt{E_n^{(+)}}}\psi_n^{(+)}(x,a), \tag{10}$$

for all $n \geq 0$ such that physically acceptable wavefunctions exist. In the deformed case considered here, this imposes that they satisfy two conditions [39]:

(i) As for conventional SE, they should be square integrable on the (finite or infinite) interval of definition (x_1, x_2) of the potentials $V_\pm(x,a)$—i.e.,

$$\int_{x_1}^{x_2} dx\,|\psi_n^{(\pm)}(x,a)|^2 < \infty. \tag{11}$$

(ii) They should ensure the Hermiticity of $H_\pm$. This amounts to the condition

$$|\psi_n^{(\pm)}(x,a))|^2 f(x) \to 0 \qquad \text{for } x \to x_1 \text{ and } x \to x_2, \tag{12}$$

which implies an additional restriction whenever $f(x) \to \infty$ for $x \to x_1$ and/or $x \to x_2$. Equations (11) and (12) ensure the self-adjointness of $H_\pm$ and guarantee the relation $(A^\pm)^\dagger = A^\mp$.

The knowledge of the eigenvalues and eigenfunctions of H_- automatically implies the same for its partner H_+ (or vice versa). However, whenever the partner potentials V_- and V_+ are similar in shape and differ only in the parameters that appear in them—i.e.,

$$V_+(x, a_0) + g(a_0) = V_-(x, a_1) + g(a_1), \tag{13}$$

where a_1 is some function of a_0 and $g(a_0)$, $g(a_1)$ do not depend on x, then the spectrum of either Hamiltonian can be derived without reference to its partner. Here we restrict ourselves to the case of translational (or additive) shape invariance—i.e., a_1 and a_0 only differ by some additive constant. Considering then a set of parameters a_i, $i = 0, 1, 2, \ldots$, and extending condition (13) to

$$V_+(x, a_i) + g(a_i) = V_-(x, a_{i+1}) + g(a_{i+1}), \qquad i = 0, 1, 2, \ldots, \tag{14}$$

we get from Equations (3) and (5) the so-called DSI condition

$$W^2(x, a_i) + \hbar f(x)\frac{dW(x, a_i)}{dx} + g(a_i) = W^2(x, a_{i+1}) - \hbar f(x)\frac{dW(x, a_{i+1})}{dx} + g(a_{i+1}). \tag{15}$$

The eigenvalues and eigenfunctions of H_- turn out to be given by

$$E_n^{(-)}(a_0) = g(a_n) - g(a_0), \qquad n = 0, 1, 2, \ldots, \tag{16}$$

$$\psi_n^{(-)}(x, a_0) \propto A^+(a_0) A^+(a_1) \ldots A^+(a_{n-1}) \psi_0^{(-)}(x, a_n), \qquad n = 0, 1, 2, \ldots, \tag{17}$$

with $\psi_0^{(-)}(x, a_n)$ as expressed in (8).

3. Deformed Shape Invariance for Superpotentials with no Explicit Dependence on $\hbar$

As in [42–46], let us assume that the additive constant that allows us to get a_{i+1} from a_i is just $\hbar$—i.e., $a_{i+1} = a_i + \hbar$. Note that, with respect to conventions used elsewhere where the system of units is such that $\hbar = 1$, this implies some parameter re-normalization. In Appendix A, we summarize the transformations that have to be carried out on the parameters and possibly the variable used in [39–41] in order to arrive at the conventions employed here.

In the present section, we will also suppose that the dependence of $W(x, a_i)$ on $\hbar$ is entirely contained in a_i, thus leaving the case of an explicit dependence of W on $\hbar$ to Section 4.

Since Equation (15) must hold for an arbitrary value of $\hbar$, we can expand it in powers of $\hbar$ and require that the coefficient of each power vanishes. It is straightforward to show that the coefficient of $\hbar$ leads to the condition

$$W\frac{\partial W}{\partial a} - f(x)\frac{\partial W}{\partial x} + \frac{1}{2}\frac{dg}{da} = 0. \tag{18}$$

Then, the coefficient of $\hbar^2$ yields

$$\frac{\partial}{\partial a}\left[W\frac{\partial W}{\partial a} - f(x)\frac{\partial W}{\partial x} + \frac{1}{2}\frac{dg}{da}\right] = 0, \tag{19}$$

which is automatically satisfied if Equation (18) is so. Finally, the coefficient of $\hbar^n$ for $n \geq 3$ gives the condition

$$\frac{(2-n)f(x)}{n!}\frac{\partial^n W}{\partial a^{n-1}\partial x} = 0, \qquad n = 3, 4, \ldots. \tag{20}$$

This set of equations is satisfied, provided

$$\frac{\partial^3 W}{\partial a^2 \partial x} = 0. \tag{21}$$

We are therefore left with two independent conditions (18) and (21). This is similar to what happens in SUSYQM [43,44], the only difference being the appearance of the deforming function $f(x)$ in the first equation.

Before giving the set of results, we shall discuss in detail two examples, a simple one and a more involved one.

3.1. Example of the Pöschl-Teller Potential

Let us consider a deforming function $f(x) = 1 + \alpha \sin^2 x$ with $-1 < \alpha \neq 0$ and $-\frac{\pi}{2} < x < \frac{\pi}{2}$, as well as a superpotential

$$W(x,a) = (1+\alpha)a\tan x, \qquad -\frac{\pi}{2} < x < \frac{\pi}{2}, \tag{22}$$

where

$$a = \frac{1}{2(1+\alpha)}[(1+\alpha)\hbar + \Delta], \qquad \Delta = \sqrt{(1+\alpha)^2\hbar^2 + 4A(A-\hbar)}, \qquad A > \hbar. \tag{23}$$

We note that this W automatically satisfies Equation (21) and that, on inserting it in Equation (18), we obtain

$$\frac{dg}{da} = 2(1+\alpha)a, \tag{24}$$

from which $g(a) = (1+\alpha)a^2$, up to some additive constant.

From Equation (3), the starting potential can be written as

$$\begin{aligned}
V_-(x,a) &= (1+\alpha)^2 a(a-\hbar)\sec^2 x - (1+\alpha)a[(1+\alpha)a - \hbar\alpha] \\
&= A(A-\hbar)\sec^2 x - \left\{ A(A-\hbar) + \frac{1}{2}\hbar[(1+\alpha)\hbar + \Delta] \right\},
\end{aligned} \tag{25}$$

and therefore corresponds to the Pöschl–Teller (PT) potential $V = A(A-\hbar)\sec^2 x$ with ground-state energy $E_0 = A(A-\hbar) + \frac{1}{2}\hbar[(1+\alpha)\hbar + \Delta]$. On the other hand, from (16), we get

$$E_n^{(-)} = g(a+n\hbar) - g(a) = \hbar^2(1+\alpha)n(n+1) + \hbar\Delta n. \tag{26}$$

The results obtained here may be compared with those derived in [40] for $\bar{V} = \bar{A}(\bar{A}-1)\sec^2 \bar{x}$, with bound-state energies $\bar{E}_n = (E_n^{(-)} + E_0)/\hbar^2 = (\bar{\lambda} + n)^2 - \bar{\alpha}(\bar{\lambda} - n^2)$, where $\bar{\lambda} = (1+\bar{\alpha})a/\hbar$ is changed into $\bar{\lambda} + 1 + \bar{\alpha}$ when going to the partner.

3.2. Example of the Radial Harmonic Oscillator Potential

Let us now consider a deforming function $f(x) = 1 + \alpha x^2$ with $\alpha > 0$ and $0 < x < \infty$, as well as a superpotential

$$W(x,a) = a\left(-\frac{1}{x} + \alpha x\right) - b\left(\frac{1}{x} + \alpha x\right), \qquad 0 < x < \infty, \tag{27}$$

where

$$a = \frac{1}{2}\left(l + 1 + \frac{1}{2}\hbar + \frac{\Delta}{2\alpha}\right), \qquad b = \frac{1}{2}\left(l + 1 - \frac{1}{2}\hbar - \frac{\Delta}{2\alpha}\right),$$

$$\Delta = \sqrt{\omega^2 + \hbar^2\alpha^2}, \qquad \omega > 0, \qquad l = 0,1,2,\dots. \tag{28}$$

Here, when going to the partner, a is assumed to change into $a + \hbar$, while b remains constant. This W automatically satisfies Equation (21) again and Equation (18) leads to

$$\frac{dg}{da} = 8\alpha a, \tag{29}$$

from which $g(a) = 4\alpha a^2$, up to some additive constant.

Equation (3) shows that the starting potential is given by

$$\begin{aligned}
V_-(x,a) &= \alpha^2(a-b)(a-b-\hbar)x^2 + \frac{(a+b)(a+b-\hbar)}{x^2} - 2\alpha(a^2 - b^2 + \hbar a) \\
&= \frac{1}{4}\omega^2 x^2 + \frac{(l+1)(l+1-\hbar)}{x^2} - \left[\left(l+1+\frac{\hbar}{2}\right)\Delta + \hbar\alpha\left(2l+2+\frac{1}{2}\hbar\right)\right]
\end{aligned} \tag{30}$$

and therefore corresponds to the radial harmonic oscillator (RHO) potential $V = \frac{1}{4}\omega^2 x^2 + \frac{(l+1)(l+1-\hbar)}{x^2}$ with ground-state energy $E_0 = \left(l+1+\frac{\hbar}{2}\right)\Delta + \hbar\alpha\left(2l+2+\frac{1}{2}\hbar\right)$. Furthermore, Equation (16) leads to

$$E_n^{(-)} = g(a+n\hbar) - g(a) = 4\hbar\alpha n\left[\left(n+\frac{1}{2}\right)\hbar + l + 1\right] + 2\hbar n\Delta. \tag{31}$$

These results are comparable with those obtained in [40] for $\bar{V} = \frac{1}{4}\bar{\omega}^2 \bar{x}^2 + \bar{l}(\bar{l}+1)/\bar{x}^2$ with bound-state energies $\bar{E}_n = E_n^{(-)} + E_0 = 2\bar{\lambda}\bar{\mu} - \bar{\alpha}\bar{\lambda} + \bar{\mu} - 4(\bar{\alpha}\bar{\lambda} - \bar{\mu})n + 4\bar{\alpha}n^2$, where $\bar{\lambda} = -(a+b)/\hbar$ and $\bar{\mu} = (a-b)\hbar\alpha$ are changed into $\bar{\lambda} - 1$ and $\bar{\mu} + \bar{\alpha}$ when going to the partner, respectively.

3.3. Lists of Results

On proceeding, as in Sections 3.1 and 3.2, we analyzed the other sets of potentials and deforming functions considered in [39–41]. They include the Scarf I (S), radial Coulomb (C), Morse (M), Eckart (E), Rosen-Morse I (RM), shifted harmonic oscillator (SHO), deformed radial harmonic oscillator (DRHO), and deformed radial Coulomb (DC) potentials. The list of them is given in Table 1 in the notations used in this paper. In all the cases, except for the PT and DC potentials, the deformed superpotential is written in terms of two combinations of parameters, the first one a being changed into $a + \hbar$ and the second one b remaining constant when going to the partner. The corresponding results are listed in Table 2. In all cases, it turns out that Equation (21) is automatically satisfied, while Equation (18) leads to the expressions of $g(a)$ listed in Table 3, together with the resulting energies $E_n^{(-)}$.

Table 1. Potentials and deforming functions.

Name	V	f
PT	$A(A-\hbar)\sec^2 x$	$1 + \alpha\sin^2 x$
	$-\frac{\pi}{2} < x < \frac{\pi}{2}, A > \hbar$	$-1 < \alpha \neq 0$
RHO	$\frac{1}{4}\omega^2 x^2 + \frac{(l+1)(l+1-\hbar)}{x^2}$	$1 + \alpha x^2$
	$0 < x < +\infty$	$\alpha > 0$
S	$[A(A-\hbar) + B^2]\sec^2 x - B(2A-\hbar)\sec x \tan x$	$1 + \alpha\sin x$
	$-\frac{\pi}{2} < x < \frac{\pi}{2}, A - \hbar > B > 0$	$0 < |\alpha| < 1$
C	$-\frac{e^2}{x} + \frac{(l+1)(l+1-\hbar)}{x^2}$	$1 + \alpha x$
	$0 < x < +\infty$	$\alpha > 0$
M	$B^2 e^{-2x} - B(2A+\hbar)e^{-x}$	$1 + \alpha e^{-x}$
	$-\infty < x < +\infty, A, B > 0$	$\alpha > 0$

Table 1. *Cont.*

Name	V	f
E	$A(A-\hbar)\operatorname{csch}^2 x - 2B\coth x$	$1 + \alpha e^{-x}\sinh x$
	$0 < x < +\infty,\ A \geq \frac{3}{2}\hbar,\ B > A^2$	$-2 < \alpha \neq 0$
RM	$A(A-\hbar)\csc^2 x + 2B\cot x$	$1 + \sin x(\alpha\cos x + \beta\sin x)$
	$0 < x < \pi,\ A \geq \frac{3}{2}\hbar$	$\frac{|\alpha|}{2} < \sqrt{1+\beta},\ \beta > -1$
SHO	$\frac{1}{4}\omega^2\left(x - \frac{2d}{\omega}\right)^2$	$1 + \alpha x^2 + 2\beta x$
	$-\infty < x < +\infty$	$\alpha > \beta^2 \geq 0$
DRHO	$\frac{\omega(\omega+2\hbar\lambda)x^2}{4(1+\lambda x^2)} + \frac{(l+1)(l+1-\hbar)}{x^2}$	$\sqrt{1+\lambda x^2}$
	$0 < x < +\infty$ if $\lambda > 0$	
	$0 < x < 1/\sqrt{|\lambda|}$ if $\lambda < 0$	
DC	$-\frac{e^2}{x}\sqrt{1+\lambda x^2} + \frac{(l+1)(l+1-\hbar)}{x^2}$	$\sqrt{1+\lambda x^2}$
	$0 < x < +\infty$ if $\lambda > 0$	
	$0 < x < 1/\sqrt{|\lambda|}$ if $\lambda < 0$	

Table 2. Superpotentials and combinations of parameters.

Name	W	Parameters
PT	$(1+\alpha)a\tan x$	$a = \frac{1}{2(1+\alpha)}\left[(1+\alpha)\hbar + \Delta\right]$
		$\Delta = \sqrt{(1+\alpha)^2\hbar^2 + 4A(A-\hbar)}$
RHO	$a\left(-\frac{1}{x} + \alpha x\right) - b\left(\frac{1}{x} + \alpha x\right)$	$a = \frac{1}{2}\left(l+1 + \frac{1}{2}\hbar + \frac{\Delta}{2\alpha}\right)$
		$b = \frac{1}{2}\left(l+1 - \frac{1}{2}\hbar - \frac{\Delta}{2\alpha}\right)$
		$\Delta = \sqrt{\omega^2 + \hbar^2\alpha^2}$
S	$a(\tan x + \alpha\sec x)$	$a = \frac{1}{2}\left(\hbar + \frac{\alpha-1}{2\alpha}\Delta_+ + \frac{\alpha+1}{2\alpha}\Delta_-\right)$
	$+b(\tan x - \alpha\sec x)$	$b = \frac{1}{4\alpha}\left[(\alpha+1)\Delta_+ + (\alpha-1)\Delta_-\right]$
		$\Delta_\pm = \sqrt{\frac{1}{4}\hbar^2(1\mp\alpha)^2 + (A\pm B)(A\pm B - \hbar)}$
C	$-\frac{a+b}{x} + \frac{2b}{a+b} - \frac{\alpha}{2}(a+b)$	$a = -\frac{1}{4}\left\{e^2 + (l+1)[\alpha(l+1-\hbar) - 4]\right\}$
		$b = \frac{1}{4}\left[e^2 + \alpha(l+1)(l+1-\hbar)\right]$
M	$-\alpha(a+b)e^{-x} - \frac{1}{2}(a+b)$	$a = -\frac{1}{4\hbar\alpha}\left\{B^2 + \alpha\left[B(2A+\hbar) - 2\hbar^2\right] - 2\hbar\Delta\right\}$
	$+\frac{2\hbar b}{\alpha(a+b)}$	$b = \frac{B}{4\hbar\alpha}\left[B + \alpha(2A+\hbar)\right]$
		$\Delta = \sqrt{4B^2 + \hbar^2\alpha^2}$
E	$-(a+b)\coth x + \frac{2\hbar b}{a+b}$	$a = \frac{1}{2\hbar}\left[-B + 2\hbar A - \frac{\alpha}{2}A(A-\hbar)\right]$
	$-\frac{\alpha}{2}(a+b)$	$b = \frac{1}{2\hbar}\left[B + \frac{\alpha}{2}A(A-\hbar)\right]$
RM	$-(a+b)\cot x + \frac{2\hbar b}{a+b}$	$a = \frac{1}{2\hbar}\left[B + 2\hbar A - \frac{\alpha}{2}A(A-\hbar)\right]$
	$-\frac{\alpha}{2}(a+b)$	$b = \frac{1}{2\hbar}\left[-B + \frac{\alpha}{2}A(A-\hbar)\right]$

Table 2. *Cont.*

Name	W	Parameters
SHO	$(a+b)(\alpha x + \beta)$	$a = \frac{\hbar}{2}\left(-\hbar\frac{\beta}{4a}\omega^2 + 1 + \frac{\Delta}{\hbar\alpha} - \frac{d}{2}\hbar\omega\right)$
	$-\frac{2b}{\hbar^2\alpha(a+b)}$	$b = \frac{1}{2}\hbar^2\omega\left(\frac{\beta}{4a}\omega + \frac{d}{2}\right)$
		$\Delta = \sqrt{\omega^2 + \hbar^2\alpha^2}$
DRHO	$a\left(-\frac{1}{x}f - \frac{\lambda x}{f}\right)$	$a = \frac{1}{2}\left(l + 1 - \frac{\omega}{2\lambda}\right)$
	$+b\left(-\frac{1}{x}f + \frac{\lambda x}{f}\right)$	$b = \frac{1}{2}\left(l + 1 + \frac{\omega}{2\lambda}\right)$
DC	$-\frac{a}{x}f + \frac{e^2}{2a}$	$a = l + 1$

Table 3. Functions $g(a)$ and bound-state energies.

Name	g	$E_n^{(-)}$
PT	$(1+\alpha)a^2$	$\hbar^2(1+\alpha)n(n+1) + \hbar\Delta n$
RHO	$4\alpha a^2$	$4\hbar\alpha n\left[\left(n + \frac{1}{2}\right)\hbar + l + 1\right] + 2\hbar n\Delta$
S	$(1-\alpha^2)a^2$	$\hbar^2(1-\alpha^2)n(n+1) + \hbar\left[(1+\alpha)\Delta_+ + (1-\alpha)\Delta_-\right]n$
	$+2(1+\alpha^2)ab$	
C	$-\frac{4b^2}{(a+b)^2}$	$\hbar n(\hbar n + 2l + 2)[e^2 - \hbar\alpha(l+1)(n+1)]$
	$-\frac{\alpha^2}{4}a(a+2b)$	$\times[e^2 + \alpha(l+1)(\hbar n + 2l + 2 - \hbar)]$
		$\times[4(l+1)^2(l+1+n\hbar)^2]^{-1}$
M	$-\frac{4\hbar^2 b^2}{\alpha^2(a+b)^2}$	$\hbar n[\Delta + \hbar\alpha(n+1)][2B(2A + \hbar) - \hbar(\Delta + \alpha\hbar)(n+1)]$
	$-\frac{1}{4}a(a+2b)$	$\times[4B^2 + 2\alpha B(2A + \hbar) + \hbar\alpha(\Delta + \hbar\alpha)(n+1)]$
		$\times(\Delta + \hbar\alpha)^{-2}[\Delta + (2n+1)\hbar\alpha]^{-2}$
E	$-\frac{4\hbar^2 b^2}{(a+b)^2}$	$\hbar n(2A + n\hbar)[4A^2(A + n\hbar)^2]^{-1}$
	$-\frac{1}{4}(\alpha + 2)^2 a(a+2b)$	$\times\{2B + 2A(A + n\hbar) + \alpha A[2A + (n-1)\hbar]\}$
		$\times[2B - 2A(A + n\hbar) - \hbar\alpha A(n+1)]$
RM	$-\frac{4\hbar^2 b^2}{(a+b)^2}$	$\hbar n(2A + n\hbar)[4A^2(A + n\hbar)^2]^{-1}$
	$-\frac{1}{4}(\alpha^2 - 4\beta - 4)a(a+2b)$	$\times\{4A^2(A + n\hbar)^2 + 4B^2 - 4\alpha BA(A - \hbar)$
		$-\hbar\alpha^2 A^2[2A - \hbar + n(2A + n\hbar)] - 4\beta A^2(A + n\hbar)^2\}$
SHO	$-\frac{4b^2}{\hbar^4\alpha^2(a+b)^2}$	$4n[\Delta + (n+1)\hbar\alpha]$
	$+(\alpha - \beta^2)a(a+2b)$	$\times\{(\Delta + \hbar\alpha)[\Delta + \hbar\alpha(2n+1)]\}^{-2}$
		$\times\{\hbar^3\alpha(\alpha - \beta^2)(n+1)[2\hbar^2\alpha^2(n+1) + \omega^2(n+2)]$
		$+d\omega^2(\beta + \hbar\alpha d) + \frac{1}{4}\hbar\omega^4$
		$+\hbar^2\Delta(\alpha - \beta^2)(n+1)[2\hbar^2\alpha^2(n+1) + \omega^2]\}$
DRHO	$-4\lambda a^2$	$2n\hbar\omega - 4\hbar\lambda n(l + 1 + \hbar n)$
DC	$-\lambda a^2 - \frac{e^4}{4a^2}$	$n\hbar(2l + 2 + n\hbar)$
		$\times\left(-\lambda + \frac{e^4}{4(l+1)^2(l+1+n\hbar)^2}\right)$

4. Deformed Shape Invariance for Superpotentials with an Explicit Dependence on $\hbar$

Let us next consider the case where the superpotential contains an explicit dependence on $\hbar$. It may then be expanded in powers of $\hbar$ as

$$W(x, a, \hbar) = \sum_{n=0}^{\infty} \hbar^n W_n(x, a). \tag{32}$$

On inserting this expression in the DSI condition (15) and proceeding as in conventional SUSYQM [43,44], we arrive at the set of relations

$$\sum_{k=0}^{n} W_k W_{n-k} + f \frac{\partial W_{n-1}}{\partial x} - \sum_{s=0}^{n} \sum_{k=0}^{s} \frac{1}{(n-s)!} \frac{\partial^{n-s}}{\partial a^{n-s}} W_k W_{s-k}$$
$$+ f \sum_{k=1}^{n} \frac{1}{(k-1)!} \frac{\partial^k}{\partial a^{k-1} \partial x} W_{n-k} - \frac{1}{n!} \frac{d^n g}{da^n} = 0, \qquad n = 1, 2, \ldots. \tag{33}$$

The latter can be rewritten as

$$2f \frac{\partial W_0}{\partial x} - \frac{\partial}{\partial a}(W_0^2 + g) = 0, \tag{34}$$

$$f \frac{\partial W_1}{\partial x} - \frac{\partial}{\partial a}(W_0 W_1) = 0, \tag{35}$$

$$2f \frac{\partial W_{n-1}}{\partial x} - \sum_{s=1}^{n-1} \sum_{k=0}^{s} \frac{1}{(n-s)!} \frac{\partial^{n-s}}{\partial a^{n-s}} W_k W_{s-k} + \frac{n-2}{n!} f \frac{\partial^n W_0}{\partial a^{n-1} \partial x}$$
$$+ f \sum_{k=2}^{n-1} \frac{1}{(k-1)!} \frac{\partial^k}{\partial a^{k-1} \partial x} W_{n-k} = 0, \qquad n = 3, 4, \ldots. \tag{36}$$

In [41], two sets of rational extensions of the DRHO potential with $\lambda < 0$ considered in Section 3, referred to as type I and type II extensions, were constructed in terms of some Jacobi polynomials of degree m. The potentials belonging to these two sets were shown to be derived from superpotentials satisfying the DSI condition. The simplest potentials, corresponding to $m = 1$, turn out to be identical and given by (after changing the parameters and the variable, as explained in Appendix A)

$$V = \frac{\omega(\omega - 2\hbar|\lambda|)x^2}{4(1 - |\lambda|x^2)} + \frac{(l+1)(l+1-\hbar)}{x^2} + 4\hbar^2 \left(\frac{\omega + 2|\lambda|(l+1-\hbar)}{[\omega - 2|\lambda|(l+1)]x^2 + 2l + 2 - \hbar} \right.$$
$$\left. - \frac{2(2l+2-\hbar)(\omega - \hbar|\lambda|)}{\{[\omega - 2|\lambda|(l+1)]x^2 + 2l + 2 - \hbar\}^2} \right), \tag{37}$$

with a corresponding superpotential

$$W = \frac{\omega x}{2\sqrt{1 - |\lambda|x^2}} - \frac{l+1}{x} \sqrt{1 - |\lambda|x^2} + 2\hbar[\omega - 2(l+1)|\lambda|]x\sqrt{1 - |\lambda|x^2}$$
$$\times \left(\frac{1}{[\omega - 2(l+1)|\lambda|]x^2 + 2l + 2 - \hbar} - \frac{1}{[\omega - 2(l+1)|\lambda|]x^2 + 2l + 2 + \hbar} \right). \tag{38}$$

Let us show that such a superpotential can be derived by the present method. For such a purpose, we plan to prove that for

$$f = \sqrt{1 - |\lambda|x^2}, \qquad a = \frac{1}{2}\left(l + 1 + \frac{\omega}{2|\lambda|}\right), \qquad b = \frac{1}{2}\left(l + 1 - \frac{\omega}{2|\lambda|}\right), \tag{39}$$

the functions

$$W_0(x, a) = -\frac{a+b}{x}f + \frac{(a-b)|\lambda|x}{f}, \tag{40}$$

$$W_{2v+1}(x, a) = 0, \qquad v = 0, 1, 2, \ldots, \tag{41}$$

$$W_{2v}(x, a) = -f\frac{16b|\lambda|x}{(4bf^2 + 2a - 2b)^{2v}}, \qquad v = 1, 2, \ldots, \tag{42}$$

provide a solution of the set of Equations (34)–(36). Note that, as in Section 3, the combinations of parameters a and b become $a + \hbar$ and b for the partner, respectively.

Let us start with Equation (34). From (40), we get

$$\frac{\partial W_0}{\partial x} = \frac{a+b}{x^2}f + \frac{(a+b)|\lambda|}{f} + \frac{(a-b)|\lambda|}{f^3}, \tag{43}$$

$$\frac{\partial W_0}{\partial a} = -\frac{1}{x}f + \frac{|\lambda|x}{f}, \tag{44}$$

from which we obtain $\frac{dg}{da} = 8a|\lambda|$ and $g = 4a^2|\lambda|$ up to some additive constant. Hence, from Equation (16),

$$E_n^{(-)} = 4\hbar|\lambda|n(n\hbar + 2a) = 4\hbar|\lambda|n\left(n\hbar + l + 1 + \frac{\omega}{2|\lambda|}\right), \tag{45}$$

in agreement with the result obtained in [41].

Equation (35) is automatically satisfied since $W_1 = 0$.

Considering next Equation (36), we note that

$$\frac{\partial^n W_0}{\partial a^{n-1}\partial x} = 0, \qquad n = 3, 4, \ldots, \tag{46}$$

and that

$$\sum_{k=0}^{s} W_k W_{s-k} = 0 \qquad \text{for odd } s. \tag{47}$$

For even s, on the other hand, we easily get

$$\sum_{k=0}^{s} W_k W_{s-k} = F_s, \tag{48}$$

with F_s defined by

$$F_s = \frac{32b|\lambda|}{(4bf^2 + 2a - 2b)^s}\{-4b(s-2)f^4 + [2a + 4b(s-2)]f^2 - a + b\}. \tag{49}$$

From this result, it is straightforward to prove that

$$\frac{\partial^{n-s}}{\partial a^{n-s}}\sum_{k=0}^{s} W_k W_{s-k} = (-2)^{n-s}\frac{(n-2)!}{(s-2)!}F_n \qquad \text{for even } s. \tag{50}$$

Equations (47) and (50) then lead to

$$\sum_{s=1}^{n-1}\sum_{k=0}^{s}\frac{1}{(n-s)!}\frac{\partial^{n-s}}{\partial a^{n-s}}W_k W_{s-k} = F_n \times \begin{cases} \frac{1}{2}(3^{n-2} - 1) & \text{for even } n, \\ -\frac{1}{2}(3^{n-2} + 1) & \text{for odd } n. \end{cases} \tag{51}$$

Furthermore, we obtain

$$2f\frac{\partial W_{n-1}}{\partial x} = \begin{cases} 0 & \text{for even } n, \\ -2F_n & \text{for odd } n, \end{cases} \tag{52}$$

as well as

$$f\sum_{k=2}^{n-1}\frac{1}{(k-1)!}\frac{\partial^k}{\partial a^{k-1}\partial x}W_{n-k} = F_n \times \begin{cases} \frac{1}{2}(3^{n-2}-1) & \text{for even } n, \\ -\frac{1}{2}(3^{n-2}-3) & \text{for odd } n. \end{cases} \tag{53}$$

On inserting Equations (46), (51), (52), and (53) in Equation (36), it is clear that the latter is satisfied, which completes the proof that Equations (40)–(42) provide a solution of Equations (34)–(36).

It now only remains to use Equations (40) and (42) in

$$W(x,a,\hbar) = \sum_{v=0}^{\infty} \hbar^{2v} W_{2v}(x,a) \tag{54}$$

to obtain

$$W(x,a,\hbar) = \frac{(a-b)|\lambda|x}{f} - \frac{a+b}{x}f - 8\hbar b|\lambda|xf\left(\frac{1}{4bf^2+2a-2b-\hbar}\right.$$
$$\left. -\frac{1}{4bf^2+2a-2b+\hbar}\right), \tag{55}$$

which, after introducing the definitions of f, a, and b, given in (39), reduces to the expression (38)—i.e., the extended superpotential found in [41].

5. Conclusions

In this paper, we have shown that the approach of Gangopadhyaya, Mallow, and their coworkers of SI potentials in conventional SUSYQM can be extended to DSI ones in DSUSYQM and we have illustrated our results by considering several examples taken from [39–41]. These include both conventional potentials, for which the corresponding superpotential has no explicit dependence on $\hbar$, and a rationally-extended one, for which there is such a dependence. In all cases, it turns out that the parameter a, which is changed into $a + \hbar$ when going to the partner potential, is a combination of the potential and deforming function parameters.

An interesting open question for future investigation would be the possibility of generalizing the method to rationally-extended potentials exhibiting an "enlarged" shape invariance, for which the partner is obtained by translating some potential parameter as well as the degree m of the polynomial arising in the denominator. Such potentials are indeed known both in conventional SUSYQM [48–51] and in DSUSYQM [41].

Funding: This research received no external funding.

Acknowledgments: This work was supported by the Fonds de la Recherche Scientifique—FNRS under Grant Number 4.45.10.08.

Conflicts of Interest: The author declares no conflict of interest.

Appendix A. Going from Previously Used Conventions to the Present Ones

In this appendix, we summarize the changes that have to be carried out to go from the conventions used in [39–41] to those of the present paper. The quantities employed in the former papers are distinguished by a bar from those used here. It is worth noting too that in [39–41], the potentials used have a nonvanishing ground-state energy and must therefore be compared with $V = V_- + E_0$, where E_0 is the shift to adjust the ground-state energy of H_- to a zero value. As a consequence, $E_n^{(-)} = E_n - E_0$ corresponds to $\bar{E}_n - \bar{E}_0$.

Pöschl-Teller potential

$$\bar{V} = \bar{A}(\bar{A} - 1)\sec^2 x, \quad \bar{f}(\bar{x}) = 1 + \bar{\alpha}\sin^2\bar{x},$$
$$\bar{A} = \frac{A}{\hbar}, \quad \bar{\alpha} = \alpha, \quad \bar{x} = x,$$
$$V = A(A - \hbar)\sec^2 x = \hbar^2 \bar{V}, \quad f(x) = 1 + \alpha\sin^2 x = \bar{f}(\bar{x}),$$
$$W = \hbar\bar{W}, \quad E_n = \hbar^2\bar{E}_n. \tag{A1}$$

Radial Harmonic Oscillator potential

$$\bar{V} = \frac{1}{4}\bar{\omega}^2\bar{x}^2 + \frac{\bar{l}(\bar{l} + 1)}{\bar{x}^2}, \quad \bar{f}(\bar{x}) = 1 + \bar{\alpha}\bar{x}^2,$$
$$\bar{\omega} = \hbar\omega, \quad \bar{l} = \frac{l + 1}{\hbar} - 1, \quad \bar{\alpha} = \hbar^2\alpha, \quad \bar{x} = \frac{x}{\hbar},$$
$$V = \frac{1}{4}\omega^2 x^2 + \frac{(l + 1)(l + 1 - \hbar)}{x^2} = \bar{V}, \quad f(x) = 1 + \alpha x^2 = \bar{f}(\bar{x}),$$
$$W = \bar{W}, \quad E_n = \bar{E}_n. \tag{A2}$$

Scarf I potential

$$\bar{V} = [\bar{A}(\bar{A} - 1)] + \bar{B}^2]\sec^2\bar{x} - \bar{B}(2\bar{A} - 1)\sec\bar{x}\tan\bar{x}, \quad \bar{f}(\bar{x}) = 1 + \bar{\alpha}\sin\bar{x},$$
$$\bar{A} = \frac{A}{\hbar}, \quad \bar{B} = \frac{B}{\hbar}, \quad \bar{\alpha} = \alpha, \quad \bar{x} = x,$$
$$V = [A(A - \hbar) + B^2]\sec^2 x - B(2A - \hbar)\sec x\tan x = \hbar^2\bar{V},$$
$$f(x) = 1 + \alpha\sin x = \bar{f}(\bar{x}), \quad W = \hbar\bar{W}, \quad E_n = \hbar^2\bar{E}_n. \tag{A3}$$

Coulomb potential

$$\bar{V} = -\frac{2\bar{Z}}{\bar{x}} + \frac{\bar{l}(\bar{l} + 1)}{\bar{x}^2}, \quad \bar{f}(\bar{x}) = 1 + \bar{\alpha}\bar{x},$$
$$\bar{Z} = \frac{e^2}{2\hbar}, \quad \bar{l} = \frac{l + 1}{\hbar} - 1, \quad \bar{\alpha} = \hbar\alpha, \quad \bar{x} = \frac{x}{\hbar},$$
$$V = -\frac{e^2}{x} + \frac{(l + 1)(l + 1 - \hbar)}{x^2} = \bar{V}, \quad f(x) = 1 + \alpha x = \bar{f}(\bar{x}),$$
$$W = \bar{W}, \quad E_n = \bar{E}_n. \tag{A4}$$

Morse potential

$$\bar{V} = \bar{B}^2 e^{-2\bar{x}} - \bar{B}(2\bar{A} + 1)e^{-\bar{x}}, \quad \bar{f}(\bar{x}) = 1 + \bar{\alpha}e^{-\bar{x}},$$
$$\bar{A} = \frac{A}{\hbar}, \quad \bar{B} = \frac{B}{\hbar}, \quad \bar{\alpha} = \alpha, \quad \bar{x} = x,$$
$$V = B^2 e^{-2x} - B(2A + \hbar)e^{-x} = \hbar^2\bar{V}, \quad f(x) = 1 + \alpha e^{-x} = \bar{f}(\bar{x}),$$
$$W = \hbar\bar{W}, \quad E_n = \hbar^2\bar{E}_n. \tag{A5}$$

Eckart potential

$$\bar{V} = \bar{A}(\bar{A} - 1)\operatorname{csch}^2\bar{x} - 2\bar{B}\coth\bar{x}, \quad \bar{f}(\bar{x}) = 1 + \bar{\alpha}e^{-\bar{x}}\sinh\bar{x},$$
$$\bar{A} = \frac{A}{\hbar}, \quad \bar{B} = \frac{B}{\hbar^2}, \quad \bar{\alpha} = \alpha, \quad \bar{x} = x,$$
$$V = A(A - \hbar)\operatorname{csch}^2 x - 2B\coth x = \hbar^2\bar{V}, \quad f(x) = 1 + \alpha e^{-x}\sinh x = \bar{f}(\bar{x}),$$
$$W = \hbar\bar{W}, \quad E_n = \hbar^2\bar{E}_n. \tag{A6}$$

Rosen-Morse I potential

$$\bar{V} = \bar{A}(\bar{A} - 1)\csc^2 \bar{x} + 2\bar{B}\cot \bar{x}, \quad \bar{f}(\bar{x}) = 1 + \sin \bar{x}(\bar{\alpha}\cos \bar{x} + \bar{\beta}\sin \bar{x}),$$

$$\bar{A} = \frac{A}{\hbar}, \quad \bar{B} = \frac{B}{\hbar^2}, \quad \bar{\alpha} = \alpha, \quad \bar{\beta} = \beta, \quad \bar{x} = x,$$

$$V = A(A - \hbar)\csc^2 x + 2B\cot x = \hbar^2\bar{V}, \quad f(x) = 1 + \sin x(\alpha\cos x + \beta\sin x) = \bar{f}(\bar{x}),$$

$$W = \hbar\bar{W}, \quad E_n = \hbar^2\bar{E}_n. \tag{A7}$$

Shifted Harmonic Oscillator potential

$$\bar{V} = \frac{1}{4}\bar{\omega}^2\left(\bar{x} - \frac{2\bar{d}}{\bar{\omega}}\right)^2, \quad \bar{f}(\bar{x}) = 1 + \bar{\alpha}\bar{x}^2 + 2\bar{\beta}\bar{x},$$

$$\bar{\omega} = \hbar\omega, \quad \bar{d} = d, \quad \bar{\alpha} = \hbar^2\alpha, \quad \bar{\beta} = \hbar\beta, \quad \bar{x} = \frac{x}{\hbar},$$

$$V = \frac{1}{4}\omega^2\left(x - \frac{2d}{\omega}\right)^2 = \bar{V}, \quad f(x) = 1 + \alpha x^2 + 2\beta x = \bar{f}(\bar{x}),$$

$$W = \bar{W}, \quad E_n = \bar{E}_n. \tag{A8}$$

Deformed Radial Harmonic Oscillator potential

$$\bar{V} = \frac{\bar{\omega}(\bar{\omega} + 2\bar{\lambda})\bar{x}^2}{4(1 + \bar{\lambda}\bar{x}^2)} + \frac{\bar{l}(\bar{l} + 1)}{\bar{x}^2}, \quad \bar{f}(\bar{x}) = \sqrt{1 + \bar{\lambda}\bar{x}^2},$$

$$\bar{\omega} = \hbar\omega, \quad \bar{l} = \frac{l+1}{\hbar} - 1, \quad \bar{\lambda} = \hbar^2\lambda, \quad \bar{x} = \frac{x}{\hbar},$$

$$V = \frac{\omega(\omega + 2\hbar\lambda)x^2}{4(1 + \lambda x^2)} + \frac{(l+1)(l+1-\hbar)}{x^2} = \bar{V}, \quad f(x) = \sqrt{1 + \lambda x^2} = \bar{f}(\bar{x}),$$

$$W = \bar{W}, \quad E_n = \bar{E}_n. \tag{A9}$$

Deformed Coulomb potential

$$\bar{V} = -\frac{\bar{Q}}{\bar{x}}\sqrt{1 + \bar{\lambda}\bar{x}^2} + \frac{\bar{l}(\bar{l} + 1)}{\bar{x}^2}, \quad \bar{f}(\bar{x}) = \sqrt{1 + \bar{\lambda}\bar{x}^2},$$

$$\bar{Q} = \frac{e^2}{\hbar}, \quad \bar{l} = \frac{l+1}{\hbar} - 1, \quad \bar{\lambda} = \hbar^2\lambda, \quad \bar{x} = \frac{x}{\hbar},$$

$$V = -\frac{e^2}{x}\sqrt{1 + \lambda x^2} + \frac{(l+1)(l+1-\hbar)}{x^2} = \bar{V}, \quad f(x) = \sqrt{1 + \lambda x^2} = \bar{f}(\bar{x}),$$

$$W = \bar{W}, \quad E_n = \bar{E}_n. \tag{A10}$$

References

1. Cooper, F.; Khare, A.; Sukhatme, U. Supersymmetry and quantum mechanics. *Phys. Rep.* **1995**, *251*, 267–385. [CrossRef]
2. Junker, G. *Supersymmetric Methods in Quantum and Statistical Physics*; Springer: Berlin, Germany, 1996.
3. Bagchi, B. *Supersymmetry in Quantum and Classical Physics*; Chapman and Hall/CRC: Boca Raton, FL, USA, 2000.
4. Gangopadhyaya, A.; Mallow, J.; Rasinariu, C. *Supersymmetric Quantum Mechanics: An Introduction*; World Scientific: Singapore, 2010.
5. Gendenshtein, L. Derivation of exact spectra of the Schrödinger equation by means of supersymmetry. *JETP Lett.* **1983**, *38*, 356–359.
6. Darboux, G. *Leçons sur la Théorie Générale des Surfaces*, 2nd ed.; Gauthier-Villars: Paris, France, 1912.

7. Schrödinger, E. A method of determining quantum-mechanical eigenvalues and eigenfunctions. *Proc. R. Ir. Acad.* **1940**, *A46*, 9–16.

8. Schrödinger, E. Further studies on solving eigenvalue problems by factorization. *Proc. R. Ir. Acad.* **1941**, *A46*, 183–206.

9. Schrödinger, E. The factorization of the hypergeometric equation. *Proc. R. Ir. Acad.* **1941**, *A47*, 53–54.

10. Infeld, L.; Hull, T.E. The factorization method. *Rev. Mod. Phys.* **1951**, *23*, 21–68. [CrossRef]

11. Gómez-Ullate, D.; Grandati, Y.; Milson, R. Extended Krein-Adler theorem for the translationally shape invariant potentials. *J. Math. Phys.* **2014**, *55*, 043510. [CrossRef]

12. Quesne, C. Exceptional orthogonal polynomials, exactly solvable potentials and supersymmetry. *J. Phys. A* **2008**, *41*, 392001. [CrossRef]

13. Bagchi, B.; Quesne, C.; Roychoudhury, R. Isospectrality of conventional and new extended potentials, second-order supersymmetry and role of $\mathcal{PT}$ symmetry. *Pramana J. Phys.* **2009**, *73*, 337–347. [CrossRef]

14. Quesne, C. Solvable rational potentials and exceptional orthogonal polynomials in supersymmetric quantum mechanics. *SIGMA* **2009**, *5*, 084. [CrossRef]

15. Odake, S.; Sasaki, R. Infinitely many shape invariant potentials and new orthogonal polynomials. *Phys. Lett. B* **2009**, *679*, 414–417. [CrossRef]

16. Gómez-Ullate, D.; Kamran, R.; Milson, R. An extended class of orthogonal polynomials defined by a Sturm-Liouville problem. *J. Math. Anal. Appl.* **2009**, *359*, 352–367. [CrossRef]

17. Kempf, A. Uncertainty relation in quantum mechanics with quantum group symmetry. *J. Math. Phys.* **1994**, *35*, 4483–4496. [CrossRef]

18. Hinrichsen, H.; Kempf, A. Maximal localization in the presence of minimal uncertainties in positions and in momenta. *J. Math. Phys.* **1996**, *37*, 2121–2137. [CrossRef]

19. Kempf, A. Non-pointlike particles in harmonic oscillators. *J. Phys. A* **1997**, *30*, 2093–2102. [CrossRef]

20. Witten, E. Reflections on the fate of spacetime. *Phys. Today* **1996**, *49*, 24–30. [CrossRef]

21. Bastard, G. *Wave Mechanics Applied to Semiconductor Heterostructures*; Editions de Physique: Les Ulis, France, 1988.

22. Weisbuch, C; Vinter, B. *Quantum Semiconductor Heterostructures*; Academic: New York, NY, USA, 1997.

23. Serra, L.; Lipparini, E. Spin response of unpolarized quantum dots. *Europhys. Lett.* **1997**, *40*, 667–672. [CrossRef]

24. Harrison, P.; Valavanis, A. *Quantum Wells, Wires and Dots: Theoretical and Computational Physics of Semiconductor Nanostructures*; Wiley: Chichester, UK, 2016.

25. Barranco, M.; Pi, M.; Gatica, S.M.; Hernández, E.S.; Navarro, J. Structure and energetics of mixed ^{4}He-^{3}He drops. *Phys. Rev. B* **1997**, *56*, 8997–9003. [CrossRef]

26. Geller, M.R.; Kohn, W. Quantum mechanics of electrons in crystals with graded composition. *Phys. Rev. Lett.* **1993**, *70*, 3103–3106. [CrossRef]

27. Arias de Saavedra, F.; Boronat, J.; Polls, A.; Fabrocini, A. Effective mass of one ^{4}He atom in liquid ^{3}He. *Phys. Rev. B* **1994**, *50*, 4248(R)–4251(R). [CrossRef]

28. Puente, A.; Serra, L.; Casas, M. Dipole excitation of Na clusters with a non-local energy density functional. *Z. Phys. D* **1994**, *31*, 283–286. [CrossRef]

29. Ring, P.; Schuck, P. *The Nuclear Many Body Problem*; Springer: New York, NY, USA, 1980.

30. Bonatsos, D.; Georgoudis, P.E.; Lenis, D.; Minkov, N.; Quesne, C. Bohr Hamiltonian with a deformation-dependent mass term for the Davidson potential. *Phys. Rev. C* **2011**, *83*, 044321. [CrossRef]

31. Willatzen, M.; Lassen B. The Ben Daniel-Duke model in general nanowire structures. *J. Phys. Condens. Matter* **2007**, *19*, 136217. [CrossRef]

32. Chamel, N. Effective mass of free neutrons in neutron star crust. *Nucl. Phys. A* **2006**, *773*, 263–278. [CrossRef]

33. Infeld, L. On a new treatment of some eigenvalue problems. *Phys. Rev.* **1941**, *59*, 737–747. [CrossRef]

34. Stevenson, A.F. Note on the "Kepler problem" in a spherical space, and the factorization method of solving eigenvalue problems. *Phys. Rev.* **1941**, *59*, 842. [CrossRef]

35. Infeld, L.; Schild, A. A note on the Kepler problem in a space of constant negative curvature. *Phys. Rev.* **1945**, *67*, 121. [CrossRef]

36. Kalnins, E.G.; Miller, W., Jr.; Pogosyan, G.S. Superintegrability and associated polynomial solutions: Euclidean space and the sphere in two dimensions. *J. Math. Phys.* **1996**, *37*, 6439–6467. [CrossRef]

37. Kalnins, E.G.; Miller, W., Jr.; Pogosyan, G.S. Superintegrability on the two-dimensional hyperboloid. *J. Math. Phys.* **1997**, *38*, 5416–5433. [CrossRef]

38. Quesne, C.; Tkachuk, V.M. Deformed algebras, position-dependent effective masses and curved spaces: An exactly solvable Coulomb problem. *J. Phys. A* **2004**, *37*, 4267–4281. [CrossRef]

39. Bagchi, B.; Banerjee, A.; Quesne, C.; Tkachuk, V.M. Deformed shape invariance and exactly solvable Hamiltonians with position-dependent effective mass. *J. Phys. A* **2005**, *38*, 2929–2945. [CrossRef]

40. Quesne, C. Point canonical transformations versus deformed shape invariance for position-dependent mass Schödinger equations. *SIGMA* **2009**, *5*, 046.

41. Quesne, C. Quantum oscillator and Kepler-Coulomb problems in curved spaces: Deformed shape invariance, point canonical transformations, and rational extensions. *J. Math. Phys.* **2016**, *57*, 102101. [CrossRef]

42. Gangopadhyaya, A.; Mallow, J.V. Generating shape invariant potentials. *Int. J. Mod. Phys. A* **2008**, *23*, 4949–4978. [CrossRef]

43. Bougie, J.; Gangopadhyaya, A.; Mallow, J.V. Generation of a complete set of additive shape-invariant potentials from an Euler equation. *Phys. Rev. Lett.* **2010**, *105*, 210402. [CrossRef]

44. Bougie, J.; Gangopadhyaya, A.; Mallow, J.; Rasinariu, C. Supersymmetric quantum mechanics and solvable models. *Symmetry* **2012**, *4*, 452–473. [CrossRef]

45. Bougie, J.; Gangopadhyaya, A.; Mallow, J.V.; Rasinariu, C. Generation of a novel exactly solvable potential. *Phys. Lett. A* **2015**, *379*, 2180–2183. [CrossRef]

46. Mallow, J.V.; Gangopadhyaya, A.; Bougie, J.; Rasinariu, C. Inter-relations between additive shape invariant superpotentials. *Phys. Lett. A* **2020**, *384*, 126129. [CrossRef]

47. Mustafa, O.; Mazharimousavi, S.H. Ordering ambiguity revisited via position dependent mass pseudo-momentum operators. *Int. J. Theor. Phys.* **2007**, *46*, 1786–1796. [CrossRef]

48. Quesne, C. Revisiting (quasi-)exactly solvable rational extensions of the Morse potential. *Int. J. Mod. Phys. A* **2012**, *27*, 1250073. [CrossRef]

49. Quesne, C. Novel enlarged shape invariance property and exactly solvable rational extensions of the Rosen-Morse II and Eckart potentials. *SIGMA* **2012**, *8*, 080. [CrossRef]

50. Grandati, Y. New rational extensions of solvable potentials with finite bound state spectrum. *Phys. Lett. A* **2012**, *376*, 2866–2872. [CrossRef]

51. Grandati, Y.; Quesne, C. Confluent chains of DBT: Enlarged shape invariance and new orthogonal polynomials. *SIGMA* **2015**, *11*, 061. [CrossRef]

Publisher's Note: MDPI stays neutral with regard to jurisdictional claims in published maps and institutional affiliations.

 *symmetry*

Article

Supersymmetric Partners of the One-Dimensional Infinite Square Well Hamiltonian

Manuel Gadella [1], José Hernández-Muñoz [2], Luis Miguel Nieto [1,*] and Carlos San Millán [1]

[1] Departamento de Física Teórica, Atómica y Óptica and IMUVA, Universidad de Valladolid, 47011 Valladolid, Spain; gadella@fta.uva.es (M.G.); carlos.san-millan@alumnos.uva.es (C.S.M.)

[2] Departamento de Física Teórica de la Materia Condensada, IFIMAC Condensed Matter Physics Center, Universidad Autónoma de Madrid, 28049 Madrid, Spain; jose.hernandezm@uam.es

* Correspondence: luismiguel.nieto.calzada@uva.es; Tel.: +34-983-423754

Abstract: We find supersymmetric partners of a family of self-adjoint operators which are self-adjoint extensions of the differential operator $-d^2/dx^2$ on $L^2[-a,a]$, $a > 0$, that is, the one dimensional infinite square well. First of all, we classify these self-adjoint extensions in terms of several choices of the parameters determining each of the extensions. There are essentially two big groups of extensions. In one, the ground state has strictly positive energy. On the other, either the ground state has zero or negative energy. In the present paper, we show that each of the extensions belonging to the first group (energy of ground state strictly positive) has an infinite sequence of supersymmetric partners, such that the ℓ-th order partner differs in one energy level from both the $(\ell - 1)$-th and the $(\ell + 1)$-th order partners. In general, the eigenvalues for each of the self-adjoint extensions of $-d^2/dx^2$ come from a transcendental equation and are all infinite. For the case under our study, we determine the eigenvalues, which are also infinite, all the extensions have a purely discrete spectrum, and their respective eigenfunctions for all of its ℓ-th supersymmetric partners of each extension.

Keywords: supersymmetric quantum mechanics; self-adjoint extensions; infinite square well; contact potentials

Citation: Gadella, M.; Hernández-Muñoz, J.; Nieto, L.M.; San Millán, C. Supersymmetric Partners of the One-Dimensional Infinite Square Well Hamiltonian. *Symmetry* **2021**, *13*, 350. https://doi.org/10.3390/sym13020350

Academic Editor: Georg Junker

Received: 1 February 2021
Accepted: 17 February 2021
Published: 21 February 2021

Publisher's Note: MDPI stays neutral with regard to jurisdictional claims in published maps and institutional affiliations.

1. Introduction

The study of one dimensional models in quantum mechanics is useful in order to gain a better understanding of the properties of quantum systems. In particular, the construction of supersymmetric (SUSY) partners of given potentials allow for an analysis of one dimensional Hamiltonians that often keep similarities with the original ones. Many studies have been done in this field and a brief account of references [1–14] only covers a small part of all previous work.

In the present paper, we intend to investigate the properties of the SUSY partners of the self-adjoint determinations of the operator $-d^2/dx^2$ on $L^2[-a,a]$, $a > 0$ and finite, with appropriate boundary conditions at the points $-a$ and a. Note that this problem is closely related to the problem of the definition of the "free" Hamiltonian on the one dimensional infinite square well potential.

From our point of view, SUSY quantum mechanics is a method that pursues the identification of the class of Hamiltonians for which their spectral problem can be algebraically solved. Traditionally, people have investigated SUSY partners of well studied exactly solvable Hamiltonians that give rise to other Hamiltonians for which the spectrum coincides with the spectrum of the original Hamiltonian except for one eigenvalue. In addition, there are several examples in which one original Hamiltonian produces an infinite chain of Hamiltonians, the first element of the chain being its SUSY partner and each of the others is a partner of the previous and the next one. Here, we explore the possibility of obtaining the whole chain of partners corresponding to self-adjoint extensions of a symmetric one dimensional Hamiltonian with equal deficiency indices. Since in our case, the variety of self-adjoint extensions is quite wide, depending on four real parameters, we have expected to find interesting new results in the field as it happened to be.

The analysis of these self-adjoint extension has been done in [15]. The task of computing the SUSY partners of all the self-adjoint determinations (also called extensions) of $-d^2/dx^2$ on $L^2[-a,a]$, their spectra and their wave functions is not trivial, although can be carried out systematically.

Although the idea of self-adjoint extensions of symmetric (or Hermitian) operators on (infinite dimensional) Hilbert spaces is not yet very popular among physicists, it is, however, possible to find recent papers on the topic [16–24]. Standard quantum mechanics textbooks refer to the one dimensional infinite square well potential or the harmonic oscillator as if they were described by a unique self-adjoint Hamiltonian, which produces a neatly calculable spectrum. The mathematical reality is much more complex and may give many more possibilities for the study of quantum mechanics systems. Let us briefly address to this problem, for which a more thorough presentation can be found in mathematical textbooks [25] as well as papers addressed to the Physics community [15].

Concerning terminology, an operator, A, on a infinitely dimensional separable Hilbert space (the Hilbert space must be infinite dimensional, since otherwise all operators are continuous and defined on the whole space. In such a case, this argumentation does not make sense. A separable Hilbert space is one with a countable orthonormal basis, which is always the case in ordinary quantum mechanics) $\mathcal{H}$ is symmetric, or equivalently Hermitian if for any pair of vectors $\varphi, \psi \in \mathcal{D}(A)$, where $\mathcal{D}(A)$ is the domain of A, which must be densely defined, one has that $\langle A\varphi|\psi\rangle = \langle \varphi|A\psi\rangle$, where $\langle -|-\rangle$ denotes the scalar product on $\mathcal{H}$. This means that the adjoint, $A^\dagger$, of A extends A, $A \prec A^\dagger$ (i.e., $\mathcal{D}(A) \subset \mathcal{D}(A^\dagger)$ and $A\psi = A^\dagger\psi$, for all $\psi \in \mathcal{D}(A)$). The deficiency indices are $n_\pm := \dim \mathrm{Ran}(A^\dagger \pm iI)$, where $\mathrm{Ran}(B)$ is the range (image space) of the operator B and I is the identity operator. A symmetric (or Hermitian) operator has self-adjoint determinations (or extensions) if and only if $n_+ = n_-$ [25]. If $n_+ = n_- = 0$, this extension is unique. On the other hand, if $n_+ = n_- \neq 0$, the number of extensions is infinite and, in the case of Hilbert spaces of functions, they usually can be determined by some matching or boundary conditions that the functions in the domain of the extensions should fulfill at some points [15,25–27].

Self-adjoint determinations of the operator $-d^2/dx^2$ defined on functions supporting whatever interval, $\mathcal{K}$, in the real line $\mathbb{R}$ are used to define the so call *contact potentials* [26,28–31]. These are perturbations of the "free operator" $H_0 = -d^2/dx^2$, which are supported on a single point $x_0 \in \mathcal{K}$. Typical examples of contact potentials are the Dirac delta $\delta(x - x_0)$ or its derivative $\delta'(x - x_0)$, which define Hamiltonians of the type $H_0 + \delta(x - x_0)$ or $H_0 + \delta'(x - x_0)$ as well defined self-adjoint operators on the Hilbert space $L^2(\mathcal{K})$ [27]. These types of perturbations may serve as a good and tractable approximation for a very localized spatial perturbation and are defined via matching conditions that must satisfy the functions on the domain of the operator at x_0. Concerning the operator $-d^2/dx^2$ on $L^2[-a,a]$, some relations have been found among the boundary conditions at the borders $-a$ and a and matching conditions defining a δ or δ' perturbation at the origin [32,33].

A comment is of relevance here. Let us consider the subspace, $\mathcal{D}_0$, of all twice differentiable square integrable functions, $\varphi(x)$, in the interval $[-a,a]$, with second derivative in $L^2[-a,a]$, verifying the boundary conditions $\varphi(-a) = \varphi(a) = \varphi'(-a) = \varphi'(a) = 0$, and a differential operator of the form

$$D = -\frac{d^2}{dx^2} + p_1(x)\frac{d}{dx} + p_2(x), \tag{1}$$

where $p_1(x)$ and $p_2(x)$ are continuous real functions (with $p_1(x)$ differentiable) on $[-a,a]$. Then D is Hermitian on $\mathcal{D}_0$ with deficiency indices $(2, 2)$. It has been proven in [34] (vol. 2, p. 90) that all self-adjoint extensions of D have a purely discrete spectrum. This is precisely the case of all the self-adjoint determinations of $-d^2/dx^2$ under our study [15]. These self-adjoint extensions are characterized by a set of four real parameters, so that one particular choice of these parameters gives a unique self-adjoint determination of $-d^2/dx^2$ on $L^2[-a,a]$ and vice-versa. Although this is much less known, a similar situation emerges in the study of the one dimensional harmonic oscillator [35].

The present article intends in the first place, to complete as far as possible, the classification of the self-adjoint extensions of $-d^2/dx^2$ on $L^2[-a,a]$ given by [15]. Once this task has been done, we intend to obtain the whole chain of SUSY partners of each of the self-adjoint extensions using standard methods already developed in the theory [1]. This kind of supersymmetry intends to construct a series of potentials (in our case one-dimensional), with an energy spectrum closely related and that can be obtained from the spectrum of the original potential. Thus, being given one of our original self-adjoint extensions and being known the solution of the spectral problem, we should be able to obtain an infinite sequence of Hamiltonians such that their spectra coincides with the spectra of the previous one except for one eigenvalue, and hence from the original one except for a finite number of energy levels. We must add that all self-adjoint extensions of $-d^2/dx^2$ on $L^2[-a,a]$ have a purely discreet spectrum with an infinite number of energy levels.

The ground state for each of these extensions either has a strictly positive, zero or negative energy. Obviously, in the latter case, this fact comes from extensions which are not definitely positive. This is somehow paradoxical, due to the form of the original operator, which is $-d^2/dx^2$. This paradox is solved in [15]. For those extensions with a ground state with strictly positive energy, we have constructed the whole sequence of its SUSY partners and have given the eigenvalues and eigenfunctions for these partners. As mentioned earlier, the set of eigenvalues for each partner comes from the set of eigenvalues of the extension from which we construct the sequence of partners.

The general formalism can also be applied to obtain a sequence of Hamiltonians when the ground state of the original self-adjoint extension of $-d^2/dx^2$ on $L^2[-a,a]$ has zero or negative energy. In this case, partner Hamiltonians may be very different from the original one in the sense that they may have a finite number of eigenvalues or simply no eigenvalues. This is due to the presence of nodes in the wave function of the ground state. Nevertheless, these partners may be obtained and classified, although this discussion is left for a future publication.

This paper is organized as follows—in Section 2 we reformulate the classification given by [15] of the self-adjoint extensions of $-d^2/dx^2$ on $L^2[-a,a]$. In Section 3, we classify these extensions in terms of some other sets of parameters, not considered in [15]. In Section 4, we construct the first SUSY partners for those extensions with positive ground level energy and give the precise form of its eigenfunctions. In Section 5, we give the complete sequence of SUSY partners for each of these extensions. We close this article with a Conclusions Section and an Appendix in which we show what the correct form for the wave functions for the energy levels should be.

2. Self-Adjoint Extensions: Determination of Their Eigenvalues

Let us go back to the differential operator $H_0 := -d^2/dx^2$ defined on $L^2[-a,a]$, $a > 0$ and with domain $\mathcal{D}_0$ as above, just before (1). On $\mathcal{D}_0$, H_0 is symmetric (Hermitian) with deficiency indices $(2,2)$ [15]. According to the von-Neumann theorem [25], H_0 admits an infinite number of self-adjoint extensions labeled by four real parameters.

The adjoint operator $H_0^\dagger$ acts as $-d^2/dx^2$ on the functions of its domain (see [36,37] for a definition of the domain of the adjoint of a given densely defined operator and its properties). If ϕ is a function of such domain, we get integrating by parts:

$$\left\langle -\frac{d^2}{dx^2}\phi,\phi \right\rangle = B(\phi,\phi) + \left\langle \phi, -\frac{d^2}{dx^2}\phi \right\rangle, \tag{2}$$

where $\langle -,- \rangle$ denotes the scalar product on $L^2[-a,a]$ and

$$B(\phi,\phi) = \phi'(a)\phi^*(a) - \phi(a)\phi'^*(a) - \phi'(-a)\phi^*(-a) + \phi(-a)\phi'^*(-a), \tag{3}$$

the prime being the derivative with respect to the variable x and the asterisk meaning complex conjugate. The self-adjoint extensions of H_0 are equal to $-d^2/dx^2$ as an operator acting on the subdomains of the domain of $H_0^\dagger$ of functions with $B(\phi,\phi) = 0$. This happens if and only if there exists a 2×2 unitary matrix U such that (see [15] and references quoted therein):

$$\begin{pmatrix} 2a\phi'(-a) - i\phi(-a) \\ 2a\phi'(a) + i\phi(a) \end{pmatrix} = U \begin{pmatrix} 2a\phi'(-a) + i\phi(-a) \\ 2a\phi'(a) - i\phi(a) \end{pmatrix}. \tag{4}$$

The set of self-adjoint extensions of H_0 is in one to one correspondence with the set of 2×2 unitary operators U. Thus, each of these extensions will be labeled by its corresponding operator as H_α. Since there is a set of four real independent parameters that characterize the set of operators U, then, the set of self-adjoint extensions of H_α is also characterized by the same parameters [15]. Each of the operators U has the following form [15]:

$$U = e^{i\psi} \begin{pmatrix} m_0 - im_3 & -m_2 - im_1 \\ m_2 - im_1 & m_0 + im_3 \end{pmatrix}. \tag{5}$$

Here, ψ and m_i, $i = 0,1,2,3$ are real parameters so that $\psi \in [0, \pi]$ and $m_0^2 + m_1^2 + m_2^2 + m_3^2 = 1$, which means that only four parameters are independent [15]. The latter relation is a consequence of unitarity: the modulus of the determinant of U must be a number of modulus one.

There are some of these extensions with a clear physical interest, which does not mean that the others are irrelevant from the physics point of view. In [15], the authors distinguish three categories of extensions:

(i) Those which preserve time reversal;
(ii) Those which preserve parity;
(iii) Those preserving positivity.

Apart from these three categories, there are some other extensions. The reason why the authors of [15] single out those extensions that preserve positivity is due to the existence of extensions with negative energies. In fact, as proven in [34] (Theorem 16, vol 2, page 44), H_α may have one (which may be doubly degenerate) or two (with no degeneration) negative energy states. All other extensions have non-negative eigenvalues and are called *positivity preserving*. Only three of these positivity preserving extensions with special simplicity are discussed in [15]. We want to determine the energy levels in this situation.

In order to obtain the energy levels for a specific self-adjoint extension, H_α, of $H_0 = -d^2/dx^2$ on $L^2[-a,a]$, we have to solve the Schrödinger equation and impose on its solutions the boundary conditions that characterize the extension. These boundary conditions are given by the (4) and (6). However as stated in [15], the determination of which operators U satisfy the positivity condition as stated before involves tedious considerations. To circumvent this difficulty, let us consider the general solution of the time independent Scrödinger equation $-d^2\phi(x)/dx^2 = E\phi(x)$, with $E = s^2/(2a)^2 \geq 0$, where $2a$ is the infinite square well width (Although the energy is given, in our notation, by $\hbar^2 E/2m$, we are calling "energy" the quantity represented by E.). This general solution is

$$\phi(x) = A \cos\left(\frac{sx}{2a}\right) + B \sin\left(\frac{sx}{2a}\right). \tag{6}$$

Here, A and B have to be fixed with two conditions: (i) $\phi(x)$ should be normalized in $L^2[-a,a]$ and (ii) $\phi(x)$ should fulfill the boundary conditions (4) and (5) so that $E \geq 0$.

Let us use (6) in relation (4) giving the general matching conditions, so as to obtain the following homogeneous linear system:

$$\left(\mathcal{L}(s) - U\mathcal{M}(s)\right)\begin{pmatrix} A \\ B \end{pmatrix} = \mathcal{N}(s)\begin{pmatrix} A \\ B \end{pmatrix} = 0, \tag{7}$$

where

$$\mathcal{L}(s) = \begin{pmatrix} s\sin\frac{s}{2} - i\cos\frac{s}{2} & s\cos\frac{s}{2} + i\sin\frac{s}{2} \\ -s\sin\frac{s}{2} + i\cos\frac{s}{2} & s\cos\frac{s}{2} + i\sin\frac{s}{2} \end{pmatrix}, \qquad \mathcal{M}(s) = \begin{pmatrix} s\sin\frac{s}{2} + i\cos\frac{s}{2} & s\cos\frac{s}{2} - i\sin\frac{s}{2} \\ -s\sin(\frac{s}{2} - i\cos\frac{s}{2} & s\cos\frac{s}{2}i\sin\frac{s}{2} \end{pmatrix}. \tag{8}$$

The eigenvalues $\lambda_\pm(s)$ of the matrix $\mathcal{N}(s)$ are given by

$$\lambda_\pm(s) = \frac{\text{Tr}(\mathcal{N}(s))}{2} \pm \sqrt{\left(\frac{\text{Tr}(\mathcal{N}(s))}{2}\right)^2 - \det(\mathcal{N}(s))}. \tag{9}$$

The trace and the determinant of $\mathcal{N}(s)$ can be easily calculated and are, respectively:

$$\text{Tr}(\mathcal{N}(s)) = e^{-\frac{1}{2}i(s-2\psi)}\left(-m_3(s+1) + im_2 e^{is}(s-1)\right), \tag{10}$$

and
$$\det(\mathcal{N}(s)) = -4ie^{i\psi}\left[(m_0 + \cos\psi)\sin s + 2s(m_1 - \cos s\sin\psi) - s^2(m_0 - \cos\psi)\sin s\right]. \tag{11}$$

To begin with, let us remark that in order to have non-trivial solutions of (7) we must have
$$\det(\mathcal{N}(s)) = 0. \tag{12}$$

Then, the set of eigenvalues of $\mathcal{N}(s)$ is given by $\text{Tr}(\mathcal{N}(s))$ and 0, as may be immediately seen from (9). The condition (12) gives a relation between the values of the energy, determined by the real parameter s, since $E = s^2/(2a)^2$, and the parameters ψ, m_0 and m_1, as in (5). In consequence, the energy levels depend on the values of these three parameters only. From (11) and (12), we obtain the following two transcendental equations (one with plus sign and the other with minus sign):

$$s\sin s = \frac{m_1 - \cos s\sin\psi}{m_0 - \cos\psi} \pm \sqrt{\left(\frac{m_1 - \cos s\sin\psi}{m_0 - \cos\psi}\right)^2 + \frac{m_0 + \cos\psi}{m_0 - \cos\psi}\sin^2 s}. \tag{13}$$

This form of the transcendental equations is quite interesting, since it will serve as an efficient estimation of the energy levels when these values cannot be exactly calculated. Otherwise, they permit to obtain exact solutions whenever they exists. Let us now summarize three of the results provided by [15], which we will need later on:

- The eigenvector (A, B) of $\mathcal{N}(s)$ with 0 eigenvalue is given by

$$A = \left[i + e^{i\psi}(im_0 + m_1 - im_2 + m_3)\right]\sin\frac{s}{2} + s\left[1 + e^{i\psi}(m_0 + im_1 + m_2 + im_3)\right]\cos\frac{s}{2}, \tag{14}$$

$$B = s\left[-1 + e^{i\psi}(m_0 + im_1 + m_2 - im_3)\right]\sin\frac{s}{2} + \left[i + e^{i\psi}(im_0 - m_1 + im_2 + m_3)\right]\cos\frac{s}{2}. \tag{15}$$

These expressions generalize similar ones published in citation [14] of our Reference [15]. We see that the eigenvector depends on all the parameters $(m_0, m_1, m_2, m_3, \psi)$.
- The extensions preserving time reversal invariance, are given by

$$m_2 = 0. \tag{16}$$

- The parity preserving extensions of H_0 are those for which the eigenfunctions $\phi(x)$ verify:
$$|\phi(x)|^2 = |\phi(-x)|^2 \implies |\phi(a)|^2 = |\phi(-a)|^2. \tag{17}$$

Parity Preserving Extensions of H_0

We are interested now in getting more information on the parity preserving extensions of H_0. Then, if we use (6) in (17) we obtain that $\mathrm{Re}(A\,B^*)\sin s = 0$. Hence, either

$$\sin s = 0 \quad \text{or} \quad \mathrm{Re}(A\,B^*) = 0. \tag{18}$$

Compare to Equation (68) in citation [14] of our Reference [15]. Taking into account the values of (A, B) given in (14) and (15) and also the fact that $\det(\mathcal{N}(s)) = 0$, the second equation of (18) implies that either $m_3 = 0$ or

$$(m_3 + \sin\psi)\sin s + 2s(m_2 + \cos s \cos\psi) + s^2(\sin\psi - m_3)\sin s = 0. \tag{19}$$

Solving this equation *as if it were a quadratic equation on s* gives a pair of transcendental equations which closely resemble Equation (13). Thus, the complete set of solutions of (18) are

$$m_3 = 0, \tag{20a}$$

$$\sin s = 0, \tag{20b}$$

$$s \sin s = \frac{m_2 + \cos s \cos\psi}{m_3 - \sin\psi} \pm \sqrt{\left(\frac{m_2 + \cos s \cos\psi}{m_3 - \sin\psi}\right)^2 + \frac{m_3 + \sin\psi}{m_3 - \sin\psi}\sin^2 s}. \tag{20c}$$

Hence, when the parity is preserved, Equation (13) holds. This happens for three different situations given by Equations (20a)–(20c). These formulas, plus (13), which derives from such a general principle as $\det(\mathcal{N}(s)) = 0$, should give the energy levels for the infinite square well with parity preserving self-adjoint extensions, H_α, of H_0.

Equation (20a) does not provide any extra information, (13) being the only relation which gives information on the energy spectrum. This parity preserving condition $m_3 = 0$ has been already used in [15], although in this paper relations (20b) and (20c) are not mentioned.

Equation (20b) obviously gives an energy spectrum of the parity preserving extensions that coincides to the spectrum given by texts in Quantum Mechanics for the extension with domain given by functions with $\phi(-a) = \phi(a) = 0$. Henceforth, we shall call this extension the *textbook extension*.

Finally, (20c) gives the energy levels for other parity preserving extensions in terms of the three parameters (ψ, m_2, m_3).

In consequence, we have eight different situations for those extension having a non-negative spectrum, including those with time reversal and parity invariance, as shown in Table 1. In the next section we will analyze some of these situations.

Table 1. List of how to obtain the possible spectra as a function of the conserved properties.

Generic Spectrum: (13)		
Time reversal invariance: (13) and (16), or $m_2 = 0$		
Parity preserving:	(13) and (20a)	$m_2 = 0$
		$m_2 \neq 0$
	(13) and (20b)	$m_2 = 0$
		$m_2 \neq 0$
	(13) and (20c)	$m_2 = 0$
		$m_2 \neq 0$

3. Spectrum of the Free Particle on a Finite Interval

One of the goals of our study is to solve the eigenvalue problem for all the self-adjoint extensions, H_α, of the operator $H_0 = -d^2/dx^2$ on $L^2[-a,a]$, which from the point of view of the physicist is the infinite square well with width $2a$. As we have already seen, there are only a few of these extensions for which we may obtain an exact solution, including the *textbook extension*. For most of these extensions the energy levels are solutions of a transcendental equation and, therefore, no explicit solutions of the eigenvalue problem for these extensions can be given.

3.1. The Angular Representation of the Self-Adjoint Extensions of H_0

Due to the relation between the parameters m_i, given by

$$m_0^2 + m_1^2 + m_2^2 + m_3^2 = 1, \tag{21}$$

a new parametric representation of the self-adjoint extensions, H_α, of $H_0 = -d^2/dx^2$ on $L^2[-a,a]$ in terms of angular variables only is possible. Apart from the variable ψ, which is already angular, so that we keep it untouched, we have three other angular variables, θ_i, $i = 0,1,2$, defined by means of the following relations:

$$m_0 = \cos\theta_1 \cos\theta_0, \quad m_1 = \cos\theta_1 \sin\theta_0, \quad m_2 = \sin\theta_1 \cos\theta_2, \quad m_3 = \sin\theta_1 \sin\theta_2. \tag{22}$$

Taking into account that Equation (13) gives the values of s, and hence the energy levels, in terms of the triplet of parameters (ψ, m_0, m_1), then, according to (22), s will depend on the angular variables $(\psi, \theta_0, \theta_1)$ only. In general, we cannot solve (13) to find $s(\psi, \theta_0, \theta_1)$ explicitly. Since (13) depends on four parameters $(s, \psi, \theta_0, \theta_1)$, we cannot represent this equation in general but, as in Figure 1, we can plot the square root of the energy (essentially s) for given values of θ_1 and $\sin\theta_0$ as a function of ψ.

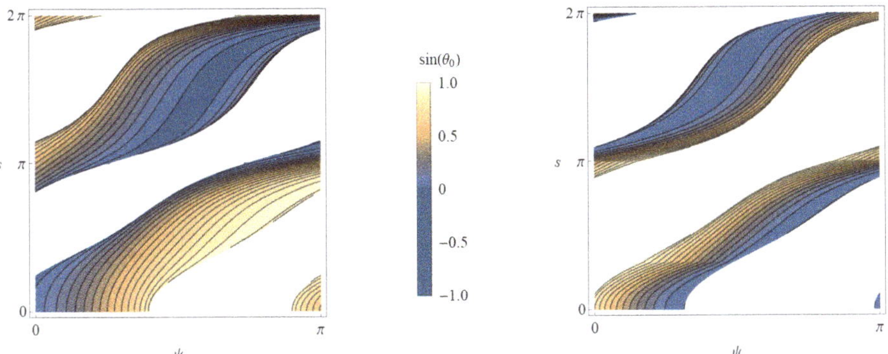

Figure 1. Two plots of the implicit Equation (13) with the parametrization (22) allow us to see the variation of the parameter s (remember that $E = s^2/(2a)^2$) as a function of ψ and $\sin\theta_0$: on the left for $\theta_1 = \pi/4$, on the right for $\theta_1 = 4\pi/3$.

The general case can neither be explicitly solved nor represented graphically. Yet, there are two other situations sharing this negative characterics. One is $m_2 = 0$ (time invariance only) and $m_3 = 0$ (parity conservation only). All other cases either can be explicitly or graphically solved or both.

3.2. Some Simple Cases

In the sequel we are going to deal with the cases of Table 1 that can be treated in some way, either graphically or analytically.

3.2.1. Parity and Time Reversal Invariance: $m_2 = m_3 = 0$

For $m_2 = m_3 = 0$, we have in (21) that $m_0^2 + m_1^2 = 1$, that is, we can take $\theta_1 = 0$ in (22) and therefore $m_0 = \cos\theta_0$ and $m_1 = \sin\theta_0$. Then, let us go back to (10), so as to see that $\mathrm{Tr}(\mathcal{N}(s)) = 0$. Since one of the eigenvalues of $\mathcal{N}(s)$ must be zero, the fact that $\mathrm{Tr}(\mathcal{N}(s)) = 0$ makes the second eigenvalue also equal to zero. Thus, the matrix $\mathcal{N}(s)$ admits a Jordan decomposition in terms of an upper triangular matrix. Now, the transcendental Equation (13) becomes much simpler, still depending on the sign, $\pm$, of the square root. Note that this sign is positive if $s \in \{(0, \pi), (2\pi, 3\pi)..\}$ and negative if $s \in \{(\pi, 2\pi), (3\pi, 4\pi),..\}$. In these two situations, the spectral equation, the vector (A, B) and the eigenfunctions have the following explicit forms:

$$
\left\{
\begin{array}{l}
\text{Positive square root }
\left\{
\begin{array}{ll}
\text{Spectrum equation:} & s\tan\left(\frac{s}{2}\right) = -\cot\left(\frac{\psi+\theta_0}{2}\right) \\
\text{Eigenvector:} & (A, B) = (1, 0) \\
\text{Eigenfunction:} & \phi(x) = \cos\left(\frac{s_0 x}{2a}\right)
\end{array}
\right. \\[3em]
\text{Negative square root }
\left\{
\begin{array}{ll}
\text{Spectrum equation:} & s\cot\left(\frac{s}{2}\right) = \cot\left(\frac{\psi-\theta_0}{2}\right) \\
\text{Eigenvector:} & (A, B) = (0, 1) \\
\text{Eigenfunction:} & \phi(x) = \sin\left(\frac{s_0 x}{2a}\right).
\end{array}
\right.
\end{array}
\right.
\tag{23}
$$

In the above expressions for the spectral equations, we may write $\frac{\psi-\theta_0}{2} = \varphi_1$ and $\frac{\psi+\theta_0}{2} = \varphi_2$, where φ_i, $i = 1, 2$ are two independent angles. Both spectral equations are represented in Figure 2. The combination of both solutions tend to the textbook solution in the limit $\varphi_{1,2} \to 0$ for both angles. The ground state for the textbook solution comes from the lowest state for the even parity preserving extensions. It is remarkable that if $\varphi_1 \geq \arctan(1/2) :=\gamma$, then the ground state no longer comes from the even but from the odd parity preserving extensions, as can be clearly seen in Figure 2.

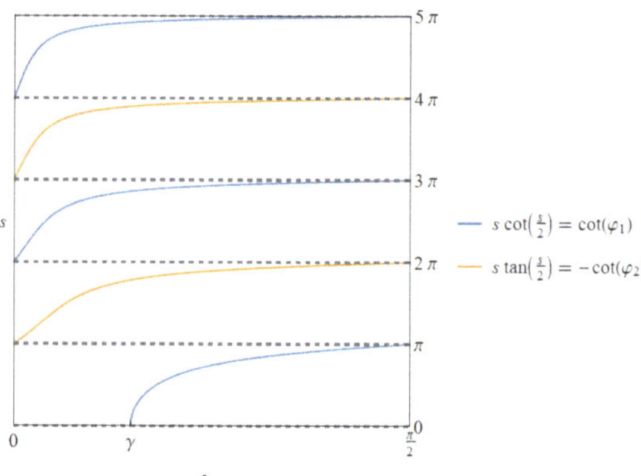

Figure 2. Energy levels ($E = s^2/(2a)^2$) for odd parity (blue) and even parity (yellow) solution for $m_2 = m_3 = 0$, coming from (23).

3.2.2. Parity Preserving Extensions Fulfilling $\sin s = 0$

In this situation, Equation (20b) implies that $s = n\pi$ with $n = 0, \pm 1, \pm 2, \ldots$, so that $E = (n^2\pi^2)/(2a)^2$, $n = 1, 2, \ldots$. The energy levels of all these extensions are the same as in the textbook's extension. No negative energy states may exist. In order to obtain the corresponding eigenfunctions, which may be different from those obtained for the textbook case, we parametrize these extensions using angular variables. However, we are now using a different angle parameterization from (22). Indeed, taking into account (13), we will find it very useful to use this one:

$$m_0 = \cos\psi\,\cos\beta_0, \qquad m_1 = (-1)^n \sin\psi, \tag{24}$$

$$m_2 = \cos\psi\,\sin\beta_0\,\cos\beta_1, \qquad m_3 = \cos\psi\,\sin\beta_0\,\sin\beta_1, \tag{25}$$

where the exponent n appearing in the expression for m_1 is the same number that labels s.

The eigenfunctions must obey the Schrödinger equation, so that they should be of the form (6). The coefficients A and B depend on the energy levels and, therefore, should be functions of s. In the simple case in which s be a even or odd multiple of 2π, these coefficients can be obtained from the following relations, using (14) and (15) and the new parameterization:

$$
\begin{cases}
s = 2q\pi: & \begin{cases} A(s) = 2\pi q\left(1 + e^{i\beta_1}\sin\beta_0 - \cos\beta_0\right), \\ B(s) = i\left(1 + e^{-i\beta_1}\sin\beta_0 + \cos\beta_0\right), \end{cases} & q = 1,2,\dots \\[2ex]
s = (2q+1)\pi: & \begin{cases} A(s) = i\left(1 - e^{i\beta_1}\sin\beta_0 + \cos\beta_0\right), \\ B(s) = \pi(2q+1)\left(-1 + e^{-i\beta_1}\sin\beta_0 + \cos\beta_0\right), \end{cases} & q = 0,1,\dots
\end{cases}
\tag{26}
$$

The eigenfunctions depend on the angles (ψ, β_0, β_1) only. These angles are those defined in (24) and (25); note the difference with the angles defined in (22). If we take the limits $\beta_0 \to 0$ and $\beta_1 \to 0$, we recover the eigenfunctions for the textbook extension.

3.2.3. Parity and Time Reversal Invariance Extensions Fulfilling (20c)

We have seen that (13) has a general validity, which is independent of the particular situation under study. On the other hand, (20c) is valid for extensions that preserve parity invariance. Note that the left hand side in both equations is the same: $s\sin(s)$. This suggests that the identity between both right hand sides would help to solve the spectral equation in this case. This identity gives:

$$
\frac{m_1 - \cos s\,\sin\psi}{m_0 - \cos\psi} \pm \sqrt{\left(\frac{m_1 - \cos s\,\sin\psi}{m_0 - \cos\psi}\right)^2 + \frac{m_0 + \cos\psi}{m_0 - \cos\psi}\sin^2 s}
$$
$$
= \frac{m_2 + \cos s\,\cos\psi}{m_3 - \sin\psi} \pm \sqrt{\left(\frac{m_2 + \cos s\,\cos\psi}{m_3 - \sin\psi}\right)^2 + \frac{m_3 + \sin\psi}{m_3 - \sin\psi}\sin^2 s}, \tag{27}
$$

which may be written in polynomial form as $\cos^4 s + P_1\cos^3 s + P_2\cos^2 s + P_3\cos s + P_4 = 0$, where the functions P_i depend on the parameters $(m_0, m_1, m_2, m_3, \psi)$. We do not write the precise form of this polynomial relation in here, since it is extremely long and it does not show interesting features. Nevertheless, it is important to note that this is a fourth order polynomial on the variable $\cos s$ with coefficients depending on the parameters. Two of these solutions of (27) are

$$\cos s = \pm 1. \tag{28}$$

These solutions may be written as $\sin s = 0$, which coincide with (20b), so that no new solutions for the spectral problem arise from (28). The other two solutions are quadratic as function of the parameters and are rather huge and intractable. To simplify this problem as much as possible, let us define a third and last angle re-parameterization of the m_i:

$$m_0 = \sin\omega_1\cos\omega_2, \quad m_1 = \cos\omega_1\sin\omega_0, \quad m_2 = \cos\omega_1\cos\omega_0, \quad m_3 = \sin\omega_1\sin\omega_2. \tag{29}$$

This parameterization is quite similar to (22), where we have interchanged the expressions for m_0 and m_2. In terms of the new angular variables, an expression for the energy levels as functions of s is given by

$$\cos s = -\cos\omega_1 \cos(\omega_0 + \omega_2) \cos(\omega_2 - \psi) \tag{30}$$

$$-\sec\omega_1 \left[\sin(\omega_0 + \omega_2)\sin(\omega_2 - \psi) \pm i\sqrt{\sin^4\omega_1 \cos^2(\omega_0 + \omega_2)\sin^2(\omega_2 - \psi)}\, \right].$$

As we want s to be real (in order to have positive eigenvalues of the energy), the imaginary term in (30) must vanish. Note that all factors under the square root are positive, so that the eigenvalues of the energy can be found, with all those in Equation (30) for each of the factors under the square root vanishing. There are three possibilities, which yield the following equations:

$$\sin\omega_1 = 0 \implies \cos s = \pm\cos(\omega_0 + \psi) \implies s = n\pi \pm (\omega_0 + \psi), \tag{31a}$$

$$\cos(\omega_0 + \omega_2) = 0 \implies \cos s = \pm\sec\omega_1 \sin(\omega_2 - \psi), \tag{31b}$$

$$\sin(\omega_2 - \psi) = 0 \implies \cos s = \pm\cos\omega_1 \cos(\omega_0 + \psi). \tag{31c}$$

Equations (31a) and (31c), give rise to an equally spaced spectrum on the variable s (not for the energy), for which $s = n\pi + f(\omega_0, \omega_1, \psi)$, $n = 0, 1, 2, \ldots$. In any case, the minimal energy level is given by $f(\omega_0, \omega_1, \psi)$. The determination of this minimal energy is not a trivial matter for (31c), since its solution $s = \arccos(\cos(\omega_1)\cos(\omega_0 + \psi))$ is given by a multi-valued function.

Equation (31b) is even more problematic, as its right hand side may be bigger than one in modulus. One may think that this formula provides the negative energy values for $|\cos s| > 1$. However, we have to keep in mind that there are only possible two negative energy levels, if any, or if there is only one, this could be either single or doubly-degenerate, so that (31b) may not give solutions to the energy spectrum and should be discarded, in principle.

3.3. About the Negative and Zero Energies

Up to now, we have not been interested in zero and negative values of the energy. Observe that the transcendental equation (13), which gives the energy levels, is valid for those extensions, H_α, having positive energies only. These energy levels are, in all positive energy cases, infinite.

If we wanted to analyze those Hamiltonians H_α with negative energy levels, we need to perform the replacement $s \to -ir$ in the wave function (6) as well as in (13). The latter appears in terms of hyperbolic functions and may have one or two solutions with zero or negative energies. If there were just one negative energy level, this is doubly degenerate [34].

When the ground state shows a negative energy, its wave function is similar to (6), where the trigonometric functions have been replaced by hyperbolic functions. In this case, the ground state wave functions may have zeros (nodes) on the interval $[-a, a]$. Here, the general formalism says that the procedure to obtain the SUSY partners is not valid [1]. Nevertheless, this formalism gives a procedure and this procedure may still be applied in this case. The result is clear—instead of obtaining a new potential with a countable infinite number of equally spaced values of the square root of the eigenvalues of the Hamiltonian (s), we obtain new Hamiltonians with either a finite number of eigenvalues or a continuous spectrum only. In the first case, these energy eigenvalues come from a transcendental equation. In the second case, partner potentials are often singular, showing an infinite divergence. We shall discuss this situation in detail in a forthcoming publication. A similar situation emerges when the ground state has zero energy.

From now on, we will concentrate in obtaining the SUSY partners of the self-adjoint extensions that we have analyzed up to now.

4. Supersymmetric Partners for the Simplest Extensions

In this section we shall consider the first and second order supersymmetry transformation applied to some of the self-adjoint extensions H_α so far considered in here.

4.1. First Order SUSY Partners

The technique to obtain the SUSY partner corresponding to a given self-adjoint operator with discrete spectrum has been discussed in [1]. To begin with, let us fix some notation and call H_α to the self-adjoint extension characterized by the values $\alpha := (m_0, m_1, m_2, m_3, \psi)$ of the parameters.

Then, let us follow the procedure of [1] to obtain the SUSY partners of H_α. First of all, we need to determine the ground state $\phi_\alpha^{(0)}(x)$ of H_α. This ground state has energy $E_\alpha^{(0)} = (s_\alpha^{(0)}/(2a))^2$, which may be in principle either positive or negative. In the present paper, we shall deal with those extensions having the ground level with positive energy and, for all energy levels, $s_\alpha^{(n)}$, $n = 0, 1, 2, \ldots$, we have $s_\alpha^{(n)} = (n+1) s_\alpha^{(0)}$.

In general, there are two supersymmetric partners of the self-adjoint extension H_α, which are Hamiltonians of the form $-d^2/dx^2 + V_\alpha^{(j)}$, where $V_\alpha^{(j)}$, $j = 1, 2$, are a pair of new potentials which is called partner potentials. In order to obtain each of the $V_\alpha^{(j)}$, pick the ground state $\phi_\alpha^{(0)}(x)$ of H_α. The explicit form of this ground state is, after (6),

$$\phi_\alpha^{(0)}(x) = A(s_0) \cos\left(\frac{s_0}{2a}x\right) + B(s_0) \sin\left(\frac{s_0}{2a}x\right), \tag{32}$$

where we have used the simplified notation $s_0 := s_\alpha^{(0)}$, which we shall henceforth keep for simplicity unless otherwise stated. Since (32) must be in the domain of H_α, the coefficients $A(s_0)$ and $B(s_0)$ must satisfy the boundary conditions defining this domain. Although these coefficients depend on the energy ground state s_0, we shall also omit this dependence, unless necessary. Then, we construct the partner potentials $V_\alpha^{(j)}$, $j = 1, 2$ using an intermediate function called the *super-potential*, $W_\alpha(x)$, which is defined as

$$W_\alpha(x) := -\frac{\partial_x \phi_\alpha^{(0)}(x)}{\phi_\alpha^{(0)}(x)}, \tag{33}$$

where ∂_x means derivative with respect to x.

Now, we construct the partner potentials $V_\alpha^{(j)}$, $j = 1, 2$, as [1]

$$V_\alpha^{(1)}(x) = W_\alpha^2(x) - W_\alpha'(x) = -\left(\frac{s_0}{2a}\right)^2, \tag{34}$$

$$V_\alpha^{(2)}(x) = W_\alpha^2(x) + W_\alpha'(x) = \left(\frac{s_0}{2a}\right)^2 \left(1 + 2\left(\frac{A \sin\left(\frac{s_0}{2a}x\right) - B \cos\left(\frac{s_0}{2a}x\right)}{A \cos\left(\frac{s_0}{2a}x\right) + B \sin\left(\frac{s_0}{2a}x\right)}\right)^2\right). \tag{35}$$

According to (34), it comes that $V_\alpha^{(1)}(x)$ is constant and equal, in modulus, to the original system lowest energy level. We see that this solution is trivial, as only shifts the energy levels. If we represent as $\phi^{(1)}(x)$ and $E_n^{(1)}$ the wave function of the ground state and the n-th energy level in this situation, we have that

$$\phi^{(1)}(x) \equiv \phi^{(0)}(x), \qquad E_n^{(1)} = \left(\frac{s_0}{2a}\right)^2 (n^2 - 1) \equiv E_n^{(0)} - E_{n=1}^{(0)}, \tag{36}$$

and $n = 1, 2, \ldots$ is arbitrary. In the sequel, we omit the subindex α for simplicity in the notation, unless otherwise stated for necessity.

The Schrödinger equation coming from the second potential in (35) is

$$-\frac{d^2}{dx^2}\phi^{(2)}(x) + \left(\frac{s_0}{2a}\right)^2\left(1 + 2\left(\frac{A\sin\left(\frac{s_0}{2a}x\right) - B\cos\left(\frac{s_0}{2a}x\right)}{A\cos\left(\frac{s_0}{2a}x\right) + B\sin\left(\frac{s_0}{2a}x\right)}\right)^2\right)\phi^{(2)}(x) = E_n^{(1)}\phi^{(2)}(x),\tag{37}$$

where the meaning of $\phi^{(2)}(x)$ is obvious. Next, let us define a new variable z as

$$\frac{s_0}{2a}z = iW(x) = \frac{is_0}{2a}\frac{A\sin\left(\frac{s_0}{2a}x\right) - B\cos\left(\frac{s_0}{2a}x\right)}{A\cos\left(\frac{s_0}{2a}x\right) + B\sin\left(\frac{s_0}{2a}x\right)}.\tag{38}$$

Under this change of variables, the Schrödinger Equation (37) takes the form:

$$(1 - z^2)\partial_z^2\phi^{(2)}(z) - 2z\,\partial_z\phi^{(2)}(z) + \left(\ell(\ell+1) - \frac{n^2}{1-z^2}\right)\phi^{(2)}(z) = 0, \quad \text{with} \quad \ell = 1, \tag{39}$$

where ∂_z represents the derivation with respect to z. This is a particular case of associated Legendre equation when $\ell = 1$, and their solutions are well known. One of them is given by the associated Legendre functions of second kind:

$$Q_\ell^n(z) := (-1)^n (1 - z^2)^{n/2}\frac{d^n}{dz^n}Q_\ell(z),\tag{40}$$

where $Q_\ell(z)$ are the Legendre functions of the second kind [38]. These solutions for (39) provide the solutions for (37):

$$\phi_n^{(2)}(x) = Q_1^n\left(i\frac{A\sin\left(\frac{s_0}{2a}x\right) - B\cos\left(\frac{s_0}{2a}x\right)}{A\cos\left(\frac{s_0}{2a}x\right) + B\sin\left(\frac{s_0}{2a}x\right)}\right),\tag{41}$$

where, of course, we have taken the value $\ell = 1$.

In addition, there is another set of solutions given by the first kind associated Legendre functions $P_1^n(z)$. These functions have not been considered as solutions to our problem, since they show singularities within the open interval $(-a, a)$ and are not square integrable, as shown in Appendix A. In Figure 3, we represent some of the wave equations just obtained in (41) for the lowest energy levels. Let us consider now the *second partner Hamiltonian*, $H_\alpha^{(2)} := H_\alpha + V_\alpha^{(2)}$, or $H^{(2)}$ in brief. The solution $Q_1^1(z)$ shows a logarithmic singularity at each of its extremes and, therefore, it is not square integrable. Nevertheless, for $n \geq 2$, these solutions are square integrable, as is proved in the Appendix A.

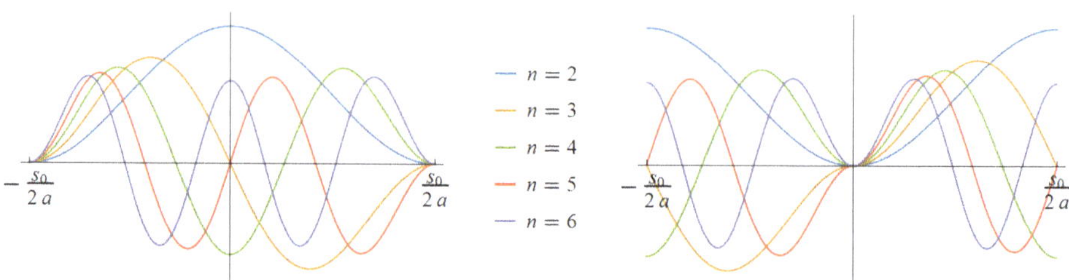

Figure 3. First order supersymmetric (SUSY) states $\phi_n^{(2)}(x)$ from (41) when the ground state of the original system is either purely even, that is $B = 0$ (plot on the left), or purely odd, that is $A = 0$ (plot on the right). Note that the quantum number n of the Legendre function in (41) is the number of the nodes of the function.

4.2. Second Order SUSY Partners

Once we have obtained the first order SUSY partners for the self-adjoint extensions H_α with ground state of positive energy, let us inspect how we may obtain an infinite chain of higher order partners for H_α. In the all above discussed cases, the bound state has wave function given by $Q_1^2(z)$ (which is obviously not the same for all cases, since the definition of z changes). Then, we obtain the super-potential $W^{(2)}(x)$ by replacing $\phi_{n=1}^{(0)}(x)$ by $Q_1^2(z)$ (and then write z in terms of x) in (33). This procedure gives rise to two second order potential partner candidates, which are:

$$V^{(2,1)}(x) = W_{(2)}^2(x) - W_{(2)}'(x) = \left(\frac{s_0}{2a}\right)^2\left(1 + 2\left(\frac{A\sin\left(\frac{s_0}{2a}x\right) - B\cos\left(\frac{s_0}{2a}x\right)}{A\cos\left(\frac{s_0}{2a}x\right) + B\sin\left(\frac{s_0}{2a}x\right)}\right)^2\right) - 3\left(\frac{s_0}{2a}\right)^2, \tag{42}$$

$$V^{(2,2)}(x) = W_{(2)}^2(x) + W_{(2)}'(x) = \left(\frac{s_0}{2a}\right)^2\left(1 + 3\left(\frac{A\sin\left(\frac{s_0}{2a}x\right) - B\cos\left(\frac{s_0}{2a}x\right)}{A\cos\left(\frac{s_0}{2a}x\right) + B\sin\left(\frac{s_0}{2a}x\right)}\right)^2\right). \tag{43}$$

Although the notation used in (42) and (43) should be clear, we need a generalization of it, as we are going to consider further order partners next. Thus, we shall use $V^{(i,j)}(x)$ and $W_{(i)}(x)$, where the index i gives the order of the partner, which in the above case is $i = 2$. This index may take all possible values $i = 1, 2, 3, \ldots$ The index j always takes two possible values, $j = 1, 2$. From this point of view, $V_\alpha^{(i)}(x)$ in (34) could be written as $V^{(1,i)}(x)$. Analogously, we may use for the i-th partner Hamiltonian the notation $H^{(i,j)}$. To simplify the notation, we have always omitted the subindex α, which labels the precise self-adjoint extension we are considering.

Observe that according to (34) and (42), $V^{(2,1)}(x) = V^{(1,2)}(x) - 3(s_0/(2a))^2 = V_\alpha^{(2)} - 3(s_0/(2a))^2$, so that $H^{(2,1)}$ and $H^{(1,2)}$ have the same eigenvalues shifted by $3(s_0/(2a))^2$. Thus, we ignore (42) and solely consider (43). For (43), we may do a similar analysis than in the previous case, that is, first order SUSY partner, so that the bound state wave functions are given by

$$\phi_n^{(2,2)}(x) = Q_2^n\left(i\frac{A\sin\left(\frac{s_0}{2a}x\right) - B\cos\left(\frac{s_0}{2a}x\right)}{A\cos\left(\frac{s_0}{2a}x\right) + B\sin\left(\frac{s_0}{2a}x\right)}\right). \tag{44}$$

In this second order SUSY, both functions $Q_2^1(z)$ and $Q_2^2(z)$ have logarithmic singularities at the points $x = \pm a$, so that they are not square integrable on $[-a, a]$ and, consequently, should be discarded as proper eigenfunctions of $H^{(2,2)}$. Thus, the ground state for $H^{(2,2)}$ has a wave function given by $Q_2^3(z)$. This is a general behaviour that could be checked at each step going from a SUSY partner to the next one, as is shown in Figure 4.

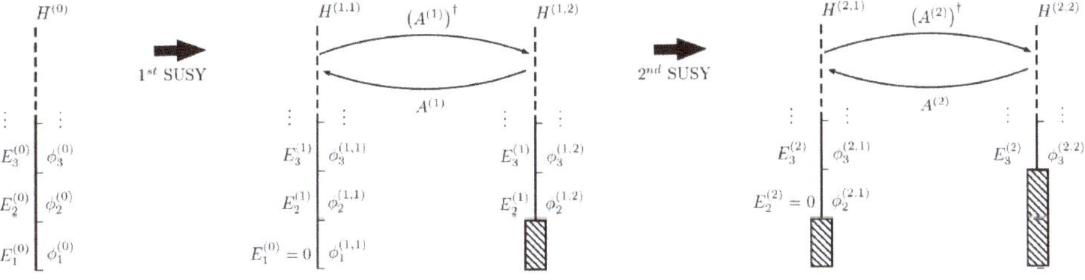

Figure 4. Different energy levels of first and second supersymmetry Hamiltonians.

5. Supersymmetric Self-Adjoint Extensions of the Infinite Well at ℓ-Order

Let us begin this Section with a summary of the notation employed so far and its meaning:

$$\begin{cases}
H_\alpha & \text{Original Hamiltonian, which is a self-adjoint extension of } H_0 = -d^2/dx^2. \\
\phi_n^{(0)} & \text{Wave function of } H_\alpha \text{ associated to the } n-\text{level.} \\
E_n^{(0)} & \text{Energy spectrum of } H_\alpha. \\
\phi_n^{(i,1)} & \text{Wave function of first SUSY partner at } i \text{ order associated to the } n- \text{ level.} \\
\phi_n^{(i,2)} & \text{Wave function of second SUSY partner at } i \text{ order associated to the } n- \text{ level.} \\
W_{(i)} & \text{Super potential at } i \text{ order, calculated from the second partner wave function.} \\
& \text{of previous SUSY order, that is, } \phi_i^{(i-1,2)}. \\
V^{(i,1)}, V^{(i,2)} & \text{Partner potentials of } i-\text{order SUSY constructed from } W_{(i)}. \\
A^{(i)}, (A^{(i)})^\dagger & \text{Annihilation/Creation operator of SUSY at } i-\text{order.}
\end{cases}$$

Creation $(A^{(i)})^\dagger$ and annihilation $A^{(i)}$ operators will be defined later.

So far, we have obtained potentials and wave functions for the first and second SUSY partners for self-adjoint extensions of $H_0 = -d^2/dx^2$ with ground level of positive energy. With the help of the induction method, we may find potentials as well as wave functions and energy levels for arbitrary order ℓ SUSY partners for the same class of self-adjoint extensions. We have seen already that from the SUSY partners $V^{(i,1)}, V^{(i,2)}$, only the last one is really interesting and we will focus on it in the sequel.

In order to apply the inductive method, let us assume that the ground state for the ℓ-th SUSY partner, $H^{(\ell,2)}$, of H_α is given by

$$\phi_{\ell+1}^{(\ell,2)}(x) = Q_\ell^{\ell+1}\left(i\frac{A\sin\left(\frac{s_0}{2a}x\right) - B\cos\left(\frac{s_0}{2a}x\right)}{A\cos\left(\frac{s_0}{2a}x\right) + B\sin\left(\frac{s_0}{2a}x\right)}\right), \tag{45}$$

as in the previous cases (41) and (44). Then, the super-potential takes the following form:

$$W_{(\ell+1)} = -\frac{\partial_x^2\phi^{(\ell,2)}(x)}{\phi^{(\ell,2)}(x)} = \frac{s_0(\ell+1)}{2a}\left(\frac{A\sin\left(\frac{s_0}{2a}x\right) - B\cos\left(\frac{s_0}{2a}x\right)}{A\cos\left(\frac{s_0}{2a}x\right) + B\sin\left(\frac{s_0}{2a}x\right)}\right) = -i\frac{s_0(\ell+1)}{2a}z, \tag{46}$$

where ∂_x^2 denotes the second derivative with respect to the variable x. Once we have the super potential, we readily obtain the partner potentials at $\ell+1$ order, which are

$$V^{(\ell+1,1)}(x) = (W_{(\ell+1)})^2 - \partial_x W_{(\ell+1)} = \frac{s_0^2(\ell+1)}{(2a)^2}\left(-1 + \ell\left(\frac{A\sin\left(\frac{s_0}{2a}x\right) - B\cos\left(\frac{s_0}{2a}x\right)}{A\cos\left(\frac{s_0}{2a}x\right) + B\sin\left(\frac{s_0}{2a}x\right)}\right)^2\right), \tag{47}$$

$$V^{(\ell+1,2)}(x) = (W_{(\ell+1)})^2 + \partial_x W_{(\ell+1)} = \frac{s_0^2(\ell+1)}{(2a)^2}\left(1 + (\ell+2)\left(\frac{A\sin\left(\frac{s_0}{2a}x\right) - B\cos\left(\frac{s_0}{2a}x\right)}{A\cos\left(\frac{s_0}{2a}x\right) + B\sin\left(\frac{s_0}{2a}x\right)}\right)^2\right). \tag{48}$$

Note that although the label α is not written explicitly on the above equations and many others, potentials and wave functions must depend on α. This dependence is hidden in s_0, where we have not made it explicitly for simplicity in the notation.

The Schrödinger equation for the first $(\ell+1)$-th order partner potential, $V^{(\ell+1,1)}$, is

$$-\partial_x^2\phi^{(\ell+1,1)}(x) + V^{(\ell+1,1)}(x)\phi^{(\ell+1,1)}(x) = E^{(\ell+1)}\phi^{(\ell+1,1)}(x). \tag{49}$$

If we change it to the z variable, (49) takes the form:

$$(1-z^2)^2\partial_z^2\phi^{(\ell+1,1)}(z) - 2z(1-z^2)\partial_z\phi^{(\ell+1,1)}(z) - (\ell+1)(1+\ell z^2)\phi^{(\ell+1,1)}(x) = \left(\frac{2a}{s_0}\right)^2 E^{(\ell+1)}\phi^{(\ell+1,1)}(x). \tag{50}$$

Equation (50) is a new Legendre-type equation for which solutions are known. The respective eigenfunctions and eigenvalues are

$$\phi_n^{(\ell+1,1)}(x) = Q_\ell^n\left(i\frac{A\sin\left(\frac{s_0}{2a}x\right) - B\cos\left(\frac{s_0}{2a}x\right)}{A\cos\left(\frac{s_0}{2a}x\right) + B\sin\left(\frac{s_0}{2a}x\right)}\right), \qquad E_n^{(\ell+1)} = \left(\frac{s_0}{2a}\right)^2\left(n^2 - (\ell+1)^2\right). \tag{51}$$

Observe that the first partner wave functions of order $\ell+1$ are the same as the second partner wave functions of order ℓ, that is, $\phi_n^{(\ell+1,1)}(x) \equiv \phi_n^{(\ell,2)}(x)$. The Schrödinger equation with potential $V^{(\ell+1,2)}(x)$ is

$$-\partial_x^2\phi^{(\ell+1,2)}(x) + V^{(\ell+1,2)}(x)\phi^{(\ell+1,2)}(x) = E_n^{(\ell+1)}\phi^{(\ell+1,2)}(x), \tag{52}$$

which in terms of the z variable becomes:

$$(1-z^2)^2\partial_z^2\phi^{(\ell+1,2)}(z) - 2z(1-z^2)\partial_z\phi^{(\ell+1,2)}(z) + (\ell+1)(1-(\ell+2)z^2)\phi^{(\ell+1,2)}(x) = \left(\frac{2a}{s_0}\right)^2 E_n^{(\ell+1)}\phi^{(\ell+1,2)}(x). \tag{53}$$

Equation (53) is again of Legendre type and its solutions in terms of eigenfunctions have the form:

$$\phi_n^{(\ell+1,2)}(x) = Q_{\ell+1}^n\left(i\frac{A\sin\left(\frac{s_0}{2a}x\right) - B\cos\left(\frac{s_0}{2a}x\right)}{A\cos\left(\frac{s_0}{2a}x\right) + B\sin\left(\frac{s_0}{2a}x\right)}\right). \tag{54}$$

The energy spectrum is given by

$$E_n^{(\ell+1)} = \left(\frac{s_0}{2a}\right)^2\left(n^2 - (\ell+1)^2\right). \tag{55}$$

Finally, one defines the annihilation, $A^{(\ell+1)}$, and creation, $(A^{(\ell+1)})^\dagger$ operators, which transform the eigenvectors of $H^{(\ell+1,1)}$ into the eigenvectors of $H^{(\ell+1,2)}$ and reciprocally, respectively, as:

$$A^{(\ell+1)} = \partial_x + W^{(\ell+1)}(x) = \frac{is_0}{2a}(1-z^2)\partial_z - i(\ell+1)\frac{s_0}{2a}z, \tag{56}$$

$$(A^{(\ell+1)})^\dagger = -\partial_x + W^{(\ell+1)}(x) = -\frac{is_0}{2a}(1-z^2)\partial_z - i(\ell+1)\frac{s_0}{2a}z. \tag{57}$$

These creation and annihilation operators have been already constructed for the general formalism of SUSY potential partners in [1].

The relation between the Hamiltonian partners $H^{(\ell,1)}$ and $H^{(\ell,2)}$ for ℓ arbitrary are shown in Figure 5. For $\ell = 0$, there is a unique Hamiltonian, which is H_α. Now, the creation and annihilation operators in the z variable give the recurrence identities for the associated Legendre functions.

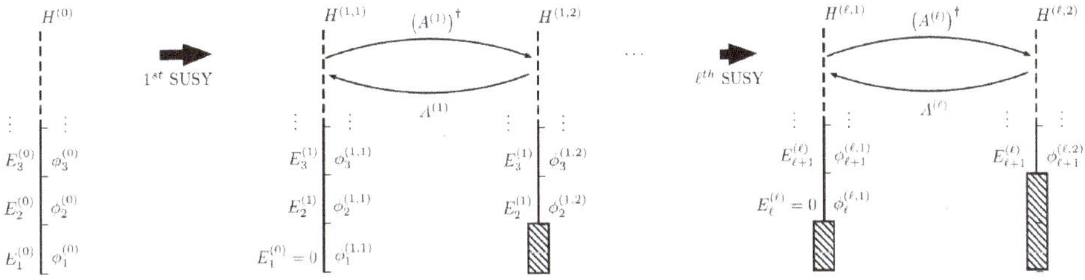

Figure 5. Energy scheme of different SUSY transformations up to order ℓ.

6. Conclusions and Outlook

We have discussed the results of [15] for the self-adjoint extensions of the differential operator $H_0 = -d^2/dx^2$ and gone beyond these results in the sense of addressing some cases not treated in [15]. Also, we have proposed a more detailed classification of the spectrum of these extensions in terms of the parameters that characterize each one of these extensions. We have seen that it is possible to classify these extensions in terms of other sets of variables with the sense of angles, which permits us to go beyond [15]. These self-adjoint extensions may have at most two negative eigenvalues, a ground state of zero energy and ground states with strictly positive energy.

In addition, in this paper, we have obtained analytically the form of the SUSY partners for the self-adjoint extensions of H_0 (that we denote as H_α, where α includes the four real parameters that gives each of these extensions) with a ground state with positive energy. We have obtained all Hamiltonian partners of each of the H_α with positive spectrum to all orders, their energy levels and their eigenfunctions. At each step, we find two distinct Hamiltonian partners of ℓ-th order. Creation and annihilation operators related the eigenfunctions for these two partners were also evaluated.

Although we have obtained the eigenfunctions for the whole sequence of SUSY partners of each of the H_α, these eigenfunctions depend explicitly on the square root of the ground state energy of H_α, which in most cases can be obtained by solving a transcendental equation. However, this transcendental equation looks rather intractable in a few cases. This situation poses some difficulties in obtaining the eigenvalues for some of the H_α, although the explicit form of their eigenfunctions and of the eigenfunctions of their SUSY partners can always be given, as functions of the square root of the ground state energy of H_α.

We have not obtained the SUSY partners for those extensions, H_α, with a ground state with zero or negative energy. Here, we may also obtain a sequence of SUSY partners form each of the H_α in this class. Unlike the partners for extensions H_α with ground states with strictly positive energies, these partners may have a finite number of eigenvalues or even none, and the potential partners may show singularities. A classification of the partners for these exceptional extensions is left for a forthcoming paper.

Author Contributions: Writing original draft: M.G., J.H.M., L.M.N. and C.S.M.; Writing review editing: M.G., J.H.M., L.M.N. and C.S.M. All authors listed have made a substantial, direct, and intellectual contribution to the work. All authors have read and agreed to the published version of the manuscript.

Funding: This research was funded by Junta de Castilla y León and FEDER projects VA137G18 and BU229P18.

Data Availability Statement: Not applicable.

Conflicts of Interest: The authors declare no conflict of interest. The funders had no role in the design of the study; in the collection, analyses, or interpretation of data; in the writing of the manuscript, or in the decision to publish the resultss.

Appendix A

In this Appendix, we justify the correct choice of the wave functions for the bound states of the supersymmetric partners of each of the extensions H_α with strictly positive ground state energy. In Appendix A.1, we derive a general solution for these wave functions as a linear combination of the associated Legendre functions P_ℓ^n and Q_ℓ^n with argument $-\tan((s_0 x)/(2a))$. In Appendix A.2, we show that the component with P_ℓ^n should be discarded, since it does not meet the requirement of square integrability. On the other hand, the component with Q_ℓ^n should give the wave function as is square integrable, as proven in Appendix A.3.

Appendix A.1

Comments in these Appendices are valid for those self-adjoint extensions H_α with ground states with positive energy. For each of these extensions, the ground state energy is $E_0^{(0)} = (s_0/(2a))^2$, where s_0 depends on the chosen self-adjoint extensions and, therefore, on the values of the parameters. As we have seen, in terms of the auxiliary variable s, the spectrum is equally spaced in this case, so that all other energy values are $E_m^{(0)} = (s_0/(2a))^2 m^2$. For the ground state, the wave function is

$$\phi_{m=1}^{(0)}(x) = A \cos\left(\frac{s_0}{2a}x\right) + B \sin\left(\frac{s_0}{2a}x\right). \tag{A1}$$

The coefficients A and B, as complex numbers, should have the same phase in order to have a real partner potential. To see it, let us write $A = Ce^{i\varphi_1}$ and $B = De^{i\varphi_2}$, with $C := |A|$ and $D := |B|$. Then, (A1) is

$$\phi_{m=1}^{(0)}(x) = Ce^{i\varphi_1} \cos\left(\frac{s_0}{2a}x\right) + De^{i\varphi_2} \sin\left(\frac{s_0}{2a}x\right). \tag{A2}$$

Using Definitions (42) and (43) for the potential partners of ℓ-th order, we have for the first ℓ-th partner:

$$V^{(\ell+1,1)} = \frac{s_0^2(\ell+1)}{(2a)^2}\left(-1 + \ell\left(\frac{Ce^{i\varphi_1}\sin\left(\frac{s_0}{2a}x\right) - De^{i\varphi_2}\cos\left(\frac{s_0}{2a}x\right)}{Ce^{i\varphi_1}\cos\left(\frac{s_0}{2a}x\right) + De^{i\varphi_2}\sin\left(\frac{s_0}{2a}x\right)}\right)^2\right), \tag{A3}$$

for which the imaginary part is given by

$$\mathrm{Im}\left(V^{(\ell+1,1)}\right) = \mathrm{Im}\left(\frac{s_0^2(\ell+1)}{(2a)^2}\left(-1 + \ell\left(\frac{A\sin\left(\frac{s_0}{2a}x\right) - B\cos\left(\frac{s_0}{2a}x\right)}{A\cos\left(\frac{s_0}{2a}x\right) + B\sin\left(\frac{s_0}{2a}x\right)}\right)^2\right)\right)$$

$$= \frac{CD\ell(\ell+1)s_0^2\left((C-D)(C+D)\sin\left(\frac{s_0 x}{a}\right) - 2CD\cos(\varphi_1-\varphi_2)\cos\left(\frac{s_0 x}{a}\right)\right)}{a^2\left(2CD\cos(\varphi_1-\varphi_2)\sin\left(\frac{s_0 x}{a}\right) + (C-D)(C+D)\cos\left(\frac{s_0 x}{a}\right) + C^2 + D^2\right)^2}\sin(\varphi_1-\varphi_2), \tag{A4}$$

so that potential (A3) is real if $\sin(\varphi_1 - \varphi_2) = 0$, or equivalently, if $\varphi_1 = n\pi + \varphi_2$. Thus, if A and B have the same phase as complex numbers, we have guaranteed that the potential partner $V^{(\ell+1,1)}$ is real. The same is valid for $V^{(\ell+1,2)}$. Thus, (A2) becomes:

$$\phi_{m=1}^{(0)}(x) = Ce^{i\varphi} \cos\left(\frac{s_0}{2a}x\right) + De^{i\varphi} \sin\left(\frac{s_0}{2a}x\right). \tag{A5}$$

This ground state is not yet normalized. Its normalization gives

$$\int_{-a}^{a} dx\, \phi_{m=1}^{(0)}(x)\left(\phi_{m=1}^{(0)}(x)\right)^* = 1 \quad \Longrightarrow \quad C^2 + D^2 = 1 \quad \Longrightarrow \quad C = \cos\delta,\ D = \sin\delta. \tag{A6}$$

Finally, the ground state wave function has the form:

$$\phi_{m=1}^{(0)}(x) = e^{i\varphi} \cos\left(\frac{s_0}{2a}x + \delta\right). \tag{A7}$$

Let us recall that our goal is to show that the solution of the Schrödinger equation with component Q_ℓ^n is square integrable and the solution with P_ℓ^n is not. To begin with, let us define a new independent variable using the shift $x = y - 2a\delta/s_0$. The ground state has now the form,

$$\phi_{m=1}^{(0)} = e^{i\varphi} \cos\left(\frac{s_0}{2a}y\right). \tag{A8}$$

With this notation, the wave function of the second partner of ℓ-th order is

$$\phi_m^{(\ell,2)} = C_1 P_\ell^n\left(-i\tan\left(\frac{s_0 y}{2a}\right)\right) + C_2 Q_\ell^n\left(-i\tan\left(\frac{s_0 y}{2a}\right)\right). \tag{A9}$$

Next, we shall analyze the square integrability of each of the components in (A9).

Appendix A.2. Trigonometric Expansion of $P_\ell^n\left(-i\tan\left(\frac{s_0 y}{2a}\right)\right)$

Let us use the change of variable $z = -i\tan\left(\frac{s_0 y}{2a}\right)$ and consider the hypergeometric form of the associated Legendre functions with argument z [38]:

$$
\begin{aligned}
P_\ell^n(z) &= \frac{1}{\ell!}\left(-\frac{1}{2}\right)^\ell\left(\frac{1+z}{1-z}\right)^{n/2}(1-z)^\ell\frac{\Gamma(2\ell+1)\,{}_2F_1\left(-\ell,n-\ell;-2\ell;-\frac{2}{z-1}\right)}{\Gamma(\ell-n+1)} \\
&= \frac{1}{\ell!}\left(-\frac{1}{2}\right)^\ell\left(\frac{1+z}{1-z}\right)^{n/2}(1-z)^\ell\sum_{j=0}^{\ell}\frac{\left((2\ell-j)!(-\ell)_j\right)}{j!\,\Gamma(-j+\ell-n+1)}\left(\frac{2}{1-z}\right)^j \\
&= \frac{\left(-\frac{1}{2}\right)^\ell\Gamma(2\ell+1)}{\ell!\,\Gamma(\ell-n+1)}\frac{e^{-\frac{is_0(n-\ell)}{2a}y}}{\cos^\ell\left(\frac{s_0 y}{2a}\right)}\sum_{j=0}^{\ell}\frac{(-1)^j\Gamma(\ell+1)(2\ell-j)!}{j!\,\Gamma(-j+\ell+1)\Gamma(-j+\ell-n+1)}\left(2e^{-\frac{is_0}{2a}y}\cos\left(\frac{s_0 y}{2a}\right)\right)^j \\
&= \sum_{j=0}^{\ell}\frac{(-1)^{j+\ell}2^{j-\ell}\Gamma(\ell+1)(2\ell-j)!}{j!\,\ell!\,\Gamma(-j+\ell+1)\Gamma(-j+\ell-n+1)}e^{-\frac{is_0 y(j-\ell+n)}{2a}}\cos^{j-\ell}\left(\frac{s_0 y}{2a}\right).
\end{aligned}
\tag{A10}
$$

Due to the presence of negative powers of the cosine in (A10), the resulting wave function is not square integrable and, therefore, not acceptable as a wave function of a bound state.

Appendix A.3. Trigonometric Expansion of $Q_\ell^n\left(-i\tan\left(\frac{s_0 y}{2a}\right)\right)$

Similarly, we can express $Q_\ell^n(z)$ in terms of a hypergeometric function [38] as:

$$
Q_\ell^n(z) = \frac{1}{\sqrt{\pi}}2^{-\ell-1}(-1)^{\ell+n+1}(z-1)^{-\ell-1}\Gamma\left(-\ell-\frac{1}{2}\right)\left(\frac{z+1}{z-1}\right)^{n/2}(\ell+n)!\,{}_2F_1\left(\ell+1,\ell+n+1;2(\ell+1);-\frac{2}{z-1}\right).
$$

Then, let us perform again the change of variables given by $z = -i\tan\left(\frac{s_0 y}{2a}\right)$, so as to obtain:

$$
\begin{aligned}
Q_\ell^n\left(-i\tan\left(\frac{s_0 y}{2a}\right)\right) &= \frac{1}{\sqrt{\pi}}2^{-\ell-1}(-1)^n\Gamma\left(-\ell-\frac{1}{2}\right)\Gamma(\ell+n+1)e^{-i\frac{s_0}{2a}y(\ell+n+1)}\cos^{\ell+1}\left(\frac{s_0 y}{2a}\right) \\
&\quad\times{}_2F_1\left(\ell+1,\ell+n+1;2(\ell+1);2e^{-i\frac{s_0 y}{2a}}\cos\left(\frac{s_0 y}{2a}\right)\right).
\end{aligned}
$$

If again, we perform a series expansion around $z = 0$ we obtain the following power series in terms of positive powers of cosines:

$$
Q_\ell^n\left(-i\tan\left(\frac{s_0 y}{2a}\right)\right) = \frac{1}{2}(-1)^{-\ell+n+1}e^{-i\frac{s_0}{2a}ny}\Gamma(n-\ell)\Gamma(\ell+n+1)\sum_{j=\ell+1}^{n}\frac{(-1)^j\Gamma(j)\,2^j e^{ij\frac{s_0}{2a}y}}{\Gamma(j-\ell)\Gamma(j+\ell+1)\Gamma(-j+n+1)}\cos^j\left(\frac{s_0}{2a}y\right).
$$

This solution is acceptable as is square integrable.

References

1. Fernández, D.J. Supersymmetric quantum mechanics. *AIP Conf. Proc.* **2010**, *1287*, 3–36.
2. Infeld, L.; Hull, T.E. The factorization method. *Rev. Mod. Phys.* **1951**, *23*, 21–68. [CrossRef]
3. Lahiri, A.; Roy, P.K.; Bagchi, B. Supersymmetry in quantum mechanics. *Int. J. Mod. Phys. A* **1990**, *5*, 1383–1456. [CrossRef]
4. Roy, B.; Roy, P.; Roychoudhury, R. On the solution of quantum eigenvalue problems. A supersymmetric point of view. *Fortschr. Phys.* **1991**, *39*, 211–258. [CrossRef]
5. Bagchi, B. *Superymmetry in Quantum and Classical Mechanics*; Chapman and Hall: Boca Raton, FL, USA, 2001.
6. Cooper, F.; Khare, A.; Sukhatme, U. Supersymmetry and Quantum Mechanics. *Phys. Rep.* **1995**, *251*, 267–385. [CrossRef]
7. Díaz, J.I.; Negro, J.; Nieto, L.M.; Rosas-Ortiz, O. The supersymmetric modified Pöschl-Teller and delta-well potential. *J. Phys. A Math. Gen.* **1999**, *32*, 8447–8460. [CrossRef]
8. Mielnik, B.; Nieto, L.M.; Rosas-Ortiz, O. The finite difference algorithm for higher order supersymmetry. *Phys. Lett. A* **2000**, *269*, 70–78. [CrossRef]
9. Fernández, C.D.J.; Negro, J.; Nieto, L.M. Regularized Scarf potentials: Energy band structure and supersymmetry. *J. Phys. A Math. Gen.* **2004**, *37*, 6987–7001.

10. Ioffe, M.V.; Negro, J.; Nieto, L.M.; Nishiniandze, D. New two-dimensional integrable quantum models from SUSY intertwining. *J. Phys. A Math. Gen.* **2006**, *39*, 9297–9308. [CrossRef]

11. Correa, F.; Nieto, L.M.; Plyushchay, M.S. Hidden nonlinear supersymmetry of finite-gap Lamé equation. *Phys. Lett. B* **2007**, *644*, 94–98. [CrossRef]

12. Ganguly, A.; Nieto, L.M. Shape-invariant quantum Hamiltonian with position-dependent effective mass through second-order supersymmetry. *J. Phys. A Math. Gen.* **2007**, *49*, 7265–7281. [CrossRef]

13. Correa, V.; Jakubsk, V.; Nieto, L.M.; Plyushchay, M.S. Self-isospectrality, special supersymmetry, and their effect on the band structure. *Phys. Rev. Lett.* **2008**, *101*, 030403. [CrossRef]

14. Fernández, C.D.J.; Gadella, M.; Nieto, L.M. Supersymmetry Transformations for Delta Potentials. *SIGMA* **2011**, *7*, 029. [CrossRef]

15. Bonneau, G.; Faraut, J.; Valent, G. Self adjoint extensions of operators and the teaching of Quantum Mechanics. *Am. J. Phys.* **2001**, *69*, 322–331. [CrossRef]

16. Albeverio, S.; Fassari, S.; Rinaldi, F. The Hamiltonian of the harmonic oscillator with an attractive δ'-interaction centred at the origin as approximated by the one with a triple of attractive delta-interactions. *J. Phys. A Math. Theor.* **2016**, *49*, 025302. [CrossRef]

17. Zolotaryuk, A.V.; Tsironis, G.P.; Zolotaryuk, Y. Point Interactions With Bias Potentials. *Front. Phys.* **2019**, *7*, 87. [CrossRef]

18. Zolotaryuk, A.V. A phenomenon of splitting resonant-tunneling one-point interactions. *Ann. Phys.* **2018**, *396*, 479–494. [CrossRef]

19. Zolotaryuk, A.V.; Zolotaryuk, Y. A zero-thickness limit of multilayer structures: A resonant-tunnelling δ'-potential. *J. Phys. A Math. Theor.* **2015**, *48*, 035302. [CrossRef]

20. Golovaty, Y.D.; Hryniv, R.O. On norm resolvent convergence of Schrödinger operators with δ'-like potentials. *J. Phys. A Math. Theor.* **2010**, *43*, 155204. [CrossRef]

21. Erman, F.; Uncu, H. Green's function formulation of multiple nonlinear Dirac delta-function potential in one dimension. *Phys. Lett. A* **2020**, *384*, 126227. [CrossRef]

22. Donaire, M.; Muñoz-Castañeda, J.M.; Nieto, L.M.; Tello-Fraile, M. Field Fluctuations and Casimir Energy of 1D-Fermions. *Symmetry* **2019**, *11*, 643. [CrossRef]

23. Gadella, M.; Mateos-Guilarte, J.M.; Muñoz-Castañeda, J.M.; Nieto, L.M. Two-point one-dimensional $\delta(x) - \delta'(x)$ interactions: Non-abelian addition law and decoupling limit. *J. Phys. A Math. Theor.* **2016**, *49*, 015204. [CrossRef]

24. Gadella, M.; Mateos-Guilarte, J.M.; Muñoz-Castañeda, J.M.; Nieto, L.M.; Santamaría-Sanz, L. Band spectra of periodic hybrid δ-δ' structures. *Eur. Phys. J. Plus* **2020**, *135*, 786. [CrossRef]

25. Reed, M.; Simon, B. *Fourier Analysis. Self Adjointness*; Academic Press: New York, NY, USA, 1975.

26. Fassari, S.; Gadella, M.; Glasser, M.L.; Nieto, L.M. Spectroscopy of a one-dimensional V-shaped quantum well with a point impurity. *Ann. Phys.* **2018**, *389*, 48–62. [CrossRef]

27. Kurasov, P. Distribution theory for discontinuous test functions and differential operators with generalized coefficients. *J. Math. Ann. Appl.* **1996**, *201*, 297–323. [CrossRef]

28. Charro, M.E.; Glasser, M.L.; Nieto, L.M. Dirac Green function for δ potentials. *EPL* **2017**, *120*, 30006. [CrossRef]

29. Fassari, S.; Gadella, M.; Nieto, L.M.; Rinaldi, F. On the spectrum of the one-dimensional Schrödinger Hamiltonian perturbed by an attractive Gaussian potential. *Acta Politech.* **2017**, *57*, 385–390. [CrossRef]

30. Muñoz-Castañeda, J.M.; Nieto, L.M.; Romaniega, C. Hyperspherical $\delta - \delta'$ potentials. *Ann. Phys.* **2019**, *400*, 246–261. [CrossRef]

31. Albeverio, S.; Fassari, S.; Gadella, M.; Nieto, L.M.; Rinaldi, F. The Birman-Schwinger Operator for a Parabolic Quantum Well in a Zero-Thickness Layer in the Presence of a Two-Dimensional Attractive Gaussian Impurity. *Front. Phys.* **2019**, *7*, 102. [CrossRef]

32. Gadella, M.; García-Ferrero, M.A.; González-Martín, S.; Maldonado-Villamizar, F.H. The infinite square well with a point interaction: A discussion on the different parameterizations. *Int. J. Theor. Phys.* **2013**, *53*, 1614–1627. [CrossRef]

33. Gadella, M.; Glasser, M.L.; Nieto, L.M. The infinite square well with a singular perturbation. *Int. J. Theor. Phys.* **2011**, *50*, 2191–2200. [CrossRef]

34. Naimark, M.A. *Linear Differential Operators*; Dover: New York, NY, USA, 2014.

35. Gadella, M.; Glasser, M.L.; Nieto, L.M. One Dimensional Models with a Singular Potential of the Type $-\alpha\,\delta(x) + \beta\delta'(x)$. *Int. J. Theor. Phys.* **2011**, *50*, 2144–2152. [CrossRef]

36. Reed, M.; Simon, B. *Functional Analysis*; Academic Press: New York, NY, USA, 1972.

37. Bachman, G.; Narici, L. *Functional Analysis*; Dover: New York, NY, USA, 2012.

38. Olver, F.W.J.; Lozier, D.W.; Boisvert, R.F.; Clark, C.W. *NIST Handbook of Mathematical Functions*, 1st ed.; Cambridge University Press: Cambridge, UK, 2010. Available online: https://dlmf.nist.gov/14.3 (accessed on 6 December 2020).

 *symmetry*

Article

Supersymmetry of Relativistic Hamiltonians for Arbitrary Spin

Georg Junker [1,2] 

[1] European Southern Observatory, Karl-Schwarzschild-Straße 2, D-85748 Garching, Germany;
   gjunker@eso.org
[2] Institut für Theoretische Physik I, Universität Erlangen-Nürnberg, Staudtstraße 7,
   D-91058 Erlangen, Germany

Received: 16 August 2020; Accepted: 22 September 2020; Published: 24 September 2020

Abstract: Hamiltonians describing the relativistic quantum dynamics of a particle with an arbitrary but fixed spin are shown to exhibit a supersymmetric structure when the even and odd elements of the Hamiltonian commute. Here, the supercharges transform between energy eigenstates of positive and negative energy. For such supersymmetric Hamiltonians, an exact Foldy–Wouthuysen transformation exists which brings it into a block-diagonal form separating the positive and negative energy subspaces. The relativistic dynamics of a charged particle in a magnetic field are considered for the case of a scalar (spin-zero) boson obeying the Klein–Gordon equation, a Dirac (spin one-half) fermion and a vector (spin-one) boson characterised by the Proca equation. In the latter case, supersymmetry implies for the Landé g-factor $g = 2$.

Keywords: relativistic wave equation; Klein–Gordon equation; Dirac equation; Proca equation; supersymmetry

1. Introduction

Soon after the formulation of non-relativistic quantum mechanics by Heisenberg, Born, Jordan and Schrödinger in 1925 and 1926, Klein [1] and Gordon [2] made first attempts to develop a relativistic quantum wave formalism. This Klein–Gordon equation is known to have certain deficits for its quantum mechanical interpretation but nowadays is well accepted as being the correct quantum wave formalism for spin-zero particles. To overcome the problems of the Klein–Gordon equation, Dirac in 1928 [3,4] made an ansatz for a wave equation being linear in the time derivative and thus found his famous equation describing the relativistic quantum dynamics of spin one-half fermions. The relativistic spin-one equation, also known as Proca equation, has been developed by Proca [5] in 1936. In the same year Dirac [6] and later Fierz and Pauli [7,8] investigated relativistic wave equations for arbitrary spin. See also the later work by Bhabha [9]. A group theoretical discussion of such wave equations was then given by Bargmann and Wigner in 1948 [10].

In their fundamental work in 1950, Foldy and Wouthuysen [11] constructed a unitary transformation which separates the positive and negative energy states; that is, the Dirac Hamiltonian became block-diagonal. This work has triggered the study of the Hamiltonian form for the other wave equations. For example, Foldy [12] investigated the Klein–Gordon equation and Feshbach and Villars [13] made a unified approach to a Hamiltonian form of the Klein–Gordon and Dirac equation. The Hamiltonian form for the Proca equation has been studied by various authors including Duffin [14], Kemmer [15], Yukawa, Sakata and Taketani [16,17], Corben and Schwinger [18], as well as by Schrödinger [19]. The problem of restoring a block-diagonal Hamiltonian via a so-called exact Foldy–Wouthuysen (FW) transformation for a particle with an arbitrary spin is still attracting many researchers. See, for example, the recent work by Silenko [20] and Simulik [21] and the references therein.

The aim of the present work was to investigate another aspect of such relativistic Hamiltonians related to an underlying supersymmetric structure. Here, it shall be emphasised that this kind of supersymmetry is not related to the supersymmetry (SUSY) known from quantum field theory where the supercharges transform a bosonic state into a fermionic state and vice versa. Here, SUSY is to be understood in the context of supersymmetric quantum mechanics, where the supercharges transform between states of positive and negative Witten parity. Despite the fact that supersymmetric quantum mechanics was originally introduced by Nicolai [22] as the $(0 + 1)$-dimensional limit of SUSY quantum field theories, it is rather independent of the latter. Supersymmetric quantum mechanics became rather popular with the model introduced by Witten [23], being in essence a one-dimensional non-relativistic quantum system, which still finds many application in various areas of physics [24,25]. The first extension of supersymmetric quantum mechanics to the relativistic Dirac Hamiltonian is due to Jackiw [26] and Ui [27] and has found many applications, for example, in the analysis of the electronic properties of topological superconductors and graphene [28].

The main purpose of the present work was to show that such supersymmetric structure may also be established for other relativistic systems going beyond that of the Dirac Hamiltonians. That is, we will show under the requirement that the odd and even part of a relativistic Hamiltonian commute with each other it is possible to establish a SUSY structure similar to what is know in the Dirac case. To the best of our knowledge, such an extension for the Klein–Gordon case was only briefly discussed by Thaller in Section 5.5.3 of his book [29] and more explicit by Znojil [30]. Here, we will present a general approach of supersymmetric quantum mechanics for relativistic Hamiltonians for arbitrary but fixed spin. The explicit discussion will be limited to the scalar or spin-zero case, the Dirac case for spin-$\frac{1}{2}$ and the vector boson case, i.e., spin-one. In all three cases we consider the well-known problem of a charged particle in the presence of a magnetic field but now from the point of view of supersymmetric quantum mechanics. It turns out that all three models, in essence, are closely related to their non-relativistic counterparts in essentially the same way. In addition, SUSY requires for vector bosons a Landé g-factor $g = 2$.

The paper is organised as follows. In the next section, the basic structure of relativistic Hamiltonians for an arbitrary spin are recalled. In Section 3, we then show that whenever the odd part commutes with the even part of such a Hamiltonian it is possible to construct an $N = 2$ SUSY structure very similar to what is known for supersymmetric Dirac Hamiltonians [29,31]. It is also recalled that there exists an exact FW transformation bringing the Hamiltonian into a block-diagonal form. In Section 4, we explicitly discuss the cases of a charged spin-zero, spin one-half and spin-one particle in an external magnetic field. In all three cases, which cover all the currently known charged elementary particles, we find that the eigenvalue problem of the relativistic Hamiltonian can indeed be reduced to that of a non-relativistic one. Section 5 discusses the resolvent of supersymmetric relativistic Hamiltonians and again shows that for the three cases under consideration the Green's function, in essence, may be reduced to that of the associated non-relativistic Hamiltonian. Finally, in Section 6, we present a short conclusion and an outlook for possible further investigations and in the Appendix A we collect some useful relations for the spin $s = 1$ case which are not that commonly known.

2. Relativistic Hamiltonians for Arbitrary Spin

In the Hamiltonian form of relativistic quantum mechanics, one puts the wave equation into a Schrödinger-like form

$$i\hbar\partial_t \Psi = H\Psi. \tag{1}$$

The Hamiltonian in the above equation in general is of the form

$$H = \beta m + \mathcal{E} + \mathcal{O}, \tag{2}$$

where $\beta^2 = 1$ acts as a grading operator and m standard for the particle's mass. In addition to the mass term βm, the operator $\mathcal{E}$ represents the remaining even part of the Hamiltonian; that is, it commutes with the grading operator, $[\beta, \mathcal{E}] = 0$. The operator $\mathcal{O}$ denotes the odd part of H and obeys the anticommutation relation $\{\beta, \mathcal{O}\} = 0$. For a particle with spin s, $s = 0, \frac{1}{2}, 1, \frac{3}{2}, \ldots$, the Hilbert space $\mathcal{H}$ on which H acts is given by

$$\mathcal{H} = L^2(\mathbb{R}^3) \otimes \mathbb{C}^{2(2s+1)}, \tag{3}$$

that is, the wave function Ψ in (1) is a spinor with $2(2s+1)$ components [20]. Let us note that we can decompose the Hilbert space $\mathcal{H}$ into a direct sum of the two eigenspaces of the grading operator β with eigenvalue $+1$ and -1, respectively,

$$\mathcal{H} = \mathcal{H}_+ \oplus \mathcal{H}_-, \qquad \mathcal{H}_\pm := L^2(\mathbb{R}^3) \otimes \mathbb{C}^{2s+1}. \tag{4}$$

Obviously, $\mathcal{H}_\pm$ are simultaneously the subspaces where eigenvalues of H are positive and negative, respectively.

For simplicity, let us put the relativistic Hamiltonian (2) into the form

$$H = \beta \mathcal{M} + \mathcal{O}, \tag{5}$$

where we have absorbed the mass m in the even mass operator $\mathcal{M} := m + \beta \mathcal{E}$ with $[\beta, \mathcal{M}] = 0$. Let us note here that above Hamiltonian is self-adjoint, i.e., $H = H^\dagger$, only for the case of fermions where $s = \frac{1}{2}, \frac{3}{2}, \ldots$ is a half-odd integer. For bosons, where s takes integer values, the Hamiltonian is pseudo-hermitian, that is, $H = \beta H^\dagger \beta$.

Choosing a representation where β takes the diagonal form

$$\beta = \begin{pmatrix} 1 & 0 \\ 0 & -1 \end{pmatrix}, \tag{6}$$

where in the above the 1 denotes a $(2s+1)$-dimensional unit matrix. The even and odd operators obeying $[\beta, \mathcal{M}] = 0$ and $\{\beta, \mathcal{O}\} = 0$ are necessarily of the form

$$\mathcal{M} = \begin{pmatrix} M_+ & 0 \\ 0 & M_- \end{pmatrix}, \qquad \mathcal{O} = \begin{pmatrix} 0 & A \\ (-1)^{2s+1} A^\dagger & 0 \end{pmatrix}, \tag{7}$$

where $M_\pm : \mathcal{H}_\pm \mapsto \mathcal{H}_\pm$ with $M_\pm^\dagger = M_\pm$ is an operator mapping positive and negative energy states into positive and negative energy states, respectively. Whereas $A : \mathcal{H}_- \mapsto \mathcal{H}_+$ maps a negative energy state into a positive energy state and $A^\dagger : \mathcal{H}_+ \mapsto \mathcal{H}_-$ vice versa. With the above representation, the general relativistic spin-s Hamiltonian then takes the form

$$H = \begin{pmatrix} M_+ & A \\ (-1)^{2s+1} A^\dagger & -M_- \end{pmatrix}. \tag{8}$$

In the following section, we will show that under the assumption that the even and odd parts of the Hamiltonian (8) commute, i.e., $[\mathcal{M}, \mathcal{O}] = 0$, it is possible to establish an $N = 2$ supersymmetric structure being well-studied in supersymmetric quantum mechanics. This condition, which we will call the SUSY condition, also allows for an exact Foldy–Wouthuysen transformation.

3. Supersymmetric Relativistic Hamiltonians for Arbitrary Spin

As stipulated above, let us assume that the even mass operator $\mathcal{M}$ and the odd operator $\mathcal{O}$ commute. This SUSY condition implies that

$$M_+ A = A M_-, \qquad A^\dagger M_+ = M_- A^\dagger. \tag{9}$$

As a consequence of this, the squared Hamiltonian (8) becomes block diagonal

$$H^2 = \begin{pmatrix} M_+^2 + (-1)^{2s+1} A A^\dagger & 0 \\ 0 & M_-^2 + (-1)^{2s+1} A^\dagger A \end{pmatrix}. \tag{10}$$

Inspired by the construction of a SUSY structure for supersymmetric Dirac Hamiltonians [31], let us introduce the following SUSY Hamiltonian

$$H_{\text{SUSY}} := \frac{(-1)^{2s+1}}{2mc^2}(H^2 - \mathcal{M}^2) = \frac{1}{2mc^2}\begin{pmatrix} A A^\dagger & 0 \\ 0 & A^\dagger A \end{pmatrix} \geq 0 \tag{11}$$

and the complex supercharges

$$Q := \frac{1}{\sqrt{2mc^2}}\begin{pmatrix} 0 & A \\ 0 & 0 \end{pmatrix}, \qquad Q^\dagger = \frac{1}{\sqrt{2mc^2}}\begin{pmatrix} 0 & 0 \\ A^\dagger & 0 \end{pmatrix}. \tag{12}$$

Here, $m > 0$ is an arbitrary mass-like parameter, representing, for example, the mass of the relativistic particle in (2). It is obvious that these operators generate a transformation between positive and negative energy states. A straightforward calculation shows that these operators together with the Witten parity operator $W := \beta$ form an $N = 2$ SUSY system; that is,

$$H_{\text{SUSY}} = \{Q, Q^\dagger\}, \qquad \{Q, W\} = 0, \qquad Q^2 = 0 = (Q^\dagger)^2, \qquad [W, H_{\text{SUSY}}] = 0, \qquad W^2 = 1. \tag{13}$$

Let us note that $\mathcal{M}$ under condition (9) commutes with all operators of above algebra and thus constitutes a centre of the SUSY algebra (13). Hence, a relativistic arbitrary-spin Hamiltonian (8) obeying the SUSY condition (9) may be called a *supersymmetric relativistic arbitrary-spin Hamiltonian*.

Let us also note that for a supersymmetric relativistic arbitrary-spin Hamiltonian, there exists an exact Foldy–Wouthuysen transformation U which brings (8) into the block-diagonal form [20]

$$\begin{aligned} H_{\text{FW}} := \quad & UHU^\dagger = \beta\sqrt{H^2} \\ = \quad & \begin{pmatrix} \sqrt{M_+^2 + (-1)^{2s+1}2mc^2 H_+} & 0 \\ 0 & -\sqrt{M_-^2 + (-1)^{2s+1}2mc^2 H_-} \end{pmatrix}, \end{aligned} \tag{14}$$

where the partner Hamiltonians $H_\pm$ are defined by

$$H_+ := \frac{1}{2mc^2} A A^\dagger, \qquad H_- := \frac{1}{2mc^2} A^\dagger A. \tag{15}$$

In fact, it is known that under condition $[\mathcal{M}, \mathcal{O}] = 0$, the exact Foldy–Wouthuysen transformation is explicitly given by [20,32]

$$U := \frac{|H| + \beta H}{\sqrt{2H^2 + 2\mathcal{M}|H|}} = \frac{1 + \beta\,\text{sgn}\,H}{\sqrt{2 + \{\text{sgn}\,H, \beta\}}}, \qquad \text{sgn}\,H := \frac{H}{\sqrt{H^2}}. \tag{16}$$

As a side remark, let us mention that the four projections operators

$$P^\pm := \tfrac{1}{2}[1 \pm W], \qquad \Lambda^\pm := \tfrac{1}{2}[1 \pm \text{sgn}\,H], \tag{17}$$

projecting onto the subspaces of positive/negative Witten parity and positive/negative eigenvalues of H, respectively, are related to each other via the same unitary transformation as H and H_{FW}

$$P^\pm = U\Lambda^\pm U^\dagger. \tag{18}$$

That is, the positive and negative energy eigenspaces are transformed via U into spaces of positive and negative Witten parity. In fact, one may verify that U may be represented in terms of these projection operators as follows

$$U = \frac{P^+\Lambda^+ + P^-\Lambda^-}{\sqrt{(P^+\Lambda^+ + P^-\Lambda^-)(\Lambda^+P^+ + \Lambda^-P^-)}}. \tag{19}$$

The non-negative partner Hamiltonians $H_\pm \geq 0$ are essential isospectral which means that their strictly positive eigenvalues are identical. The corresponding eigenstates are related to each other via a SUSY transformation. To be more explicit, let us assume these are given by

$$H_\pm \phi_\varepsilon^\pm = \varepsilon \phi_\varepsilon^\pm, \qquad \phi_\varepsilon^\pm \in \mathcal{H}_\pm, \qquad \varepsilon > 0, \tag{20}$$

then the SUSY transformation reads [31]

$$\phi_\varepsilon^+ = \frac{1}{\sqrt{2mc^2\varepsilon}} A \phi_\varepsilon^-, \qquad \phi_\varepsilon^- = \frac{1}{\sqrt{2mc^2\varepsilon}} A^\dagger \phi_\varepsilon^+. \tag{21}$$

Note that the energy eigenvalue ε may be degenerate and above relations are valid for each of these energy eigenstates. We omit an additional index in $\phi_\varepsilon^\pm$ enumerating such a degeneracy. In addition, both partner Hamiltonians $H_\pm$ may have a non-trivial kernel; that is, there may exist one or several eigenstates with

$$H_\pm \phi_0^\pm = 0. \tag{22}$$

In this case, SUSY is said to be unbroken [31]. For these ground states, again we omit the index for a possible degeneracy. There exists no SUSY transformation relating ϕ_0^+ and ϕ_0^-. The breaking of SUSY can be studied via the so-called Witten index Δ [33], which in the current context is identical to the Fredholm index of A, if it is a Fredholm operator, that is,

$$\Delta \equiv \mathrm{ind}\, A := \dim \ker A - \dim \ker A^\dagger = \dim \ker H_- - \dim \ker H_+. \tag{23}$$

Obviously a non-vanishing Witten index indicates that SUSY is unbroken. In connection with [29]

$$\dim \ker (Q + Q^\dagger) = \dim \ker A + \dim \ker A^\dagger = \dim \ker H_- + \dim \ker H_+ \tag{24}$$

the kernels of $H_\pm$; that is, the number of zero-energy states of $H_\pm$ are known. In general, however, the operator A is not Fredholm and hence some regularized indices are studied [29,31].

Due to the SUSY condition (9), the mass operators commute with the associated partner Hamiltonians, $[M_\pm, H_\pm] = 0$, and therefore have an identical set of eigenstates. Let us denote the corresponding eigenvalues of $M_\pm$ by $m_\pm c^2$; that is,

$$M_\pm \phi_\varepsilon^\pm = m_\pm c^2 \phi_\varepsilon^\pm, \qquad \varepsilon \geq 0, \tag{25}$$

then obviously the eigenvalues and eigenstates of (14),

$$H_{\mathrm{FW}} \psi_\varepsilon^\pm = E_\pm \psi_\varepsilon^\pm, \tag{26}$$

are given by

$$E_\pm = \pm\sqrt{m_\pm^2 c^4 + (-1)^{2s+1}2mc^2\varepsilon}, \qquad \psi_\varepsilon^+ = \begin{pmatrix} \phi_\varepsilon^+ \\ 0 \end{pmatrix}, \qquad \psi_\varepsilon^- = \begin{pmatrix} 0 \\ \phi_\varepsilon^- \end{pmatrix}. \tag{27}$$

Here, let us note that the mass eigenvalues may depend on the energy eigenvalues, $m_\pm = m_\pm(\varepsilon)$. Using the relation (9) in combination with the SUSY relation (21) and the above eigenvalue

Equation (25), one may verify that $m_+(\varepsilon) = m_-(\varepsilon)$ for all $\varepsilon > 0$. In essence, this means that the spectrum of a supersymmetric relativistic Hamiltonian is symmetric about zero with a possible exception at $\pm m_\pm(0)c^2$; that is, if $\varepsilon = 0$, which may only occur in the case of unbroken SUSY.

The eigenstates of the original Hamiltonian (8) are then easily found via the unitary transformation $\Psi^\pm_\varepsilon = U^\dagger \psi^\pm_\varepsilon$ having the same eigenvalues $E_\pm$ given above. Hence, the eigenvalue problem of a supersymmetric relativistic spin-s Hamiltonian can be reduced to the simultaneous eigenvalue problems for $M_\pm$ and $H_\pm$ on $\mathcal{H}_\pm$.

It will turn out in the examples to be discussed below that the partner Hamiltonians $H_\pm$ and the mass operators $M_\pm$ are in essence represented by a non-relativistic Schrödinger-like Hamiltonian H_{NR} and/or some constant operator. To be more precise, we will show for all three cases—$s = 0, \frac{1}{2}$ and 1—discussed below that the FW-transformed relativistic Hamiltonian takes the form

$$H_{\mathrm{FW}} = \beta mc^2 \sqrt{1 + \frac{2H_{\mathrm{NR}}}{mc^2}} \tag{28}$$

with H_{NR} representing the associate non-relativistic Hamiltonian as

$$\lim_{c \to \infty} \left(P^\pm H_{\mathrm{FW}} P^\pm \mp mc^2 \right) = \pm \lim_{c \to \infty} \left(mc^2 \sqrt{1 + 2H_{\mathrm{NR}}/mc^2} - mc^2 \right) = \pm H_{\mathrm{NR}} . \tag{29}$$

4. Examples

In the following subsections we will consider a relativistic charged particle with charge e and mass m in an external magnetic field $\vec{B} := \vec{\nabla} \times \vec{A}$ characterised by a vector potential $\vec{A}$. The symbol $\vec{\pi}$ stands for the kinetic momentum; that is, $\vec{\pi} := \vec{p} - e\vec{A}/c$, where c denotes the speed of light. We will consider the case of a scalar particle with spin $s = 0$, a Dirac particle with $s = \frac{1}{2}$ and a vector boson having spin $s = 1$.

4.1. The Klein–Gordon Hamiltonian with Magnetic Field

The Schrödinger form of the Klein–Gordon equation as been considered by Feshbach and Villars in [13] where they have shown that the Klein–Gordon equation for a charged particle in a magnetic field can be put into the Schrödinger form (1) where the pseudo-Hermitian Hamiltonian is given by

$$H = \begin{pmatrix} mc^2 + \frac{\vec{\pi}^2}{2m} & \frac{\vec{\pi}^2}{2m} \\ -\frac{\vec{\pi}^2}{2m} & -mc^2 - \frac{\vec{\pi}^2}{2m} \end{pmatrix} , \qquad \mathcal{H} = L^2(\mathbb{R}^3) \otimes \mathbb{C}^2 . \tag{30}$$

Comparing this with the general form (8), we may identify the operators $M_\pm$ and A as follows:

$$M_\pm = \frac{1}{2m}\vec{\pi}^2 + mc^2 , \qquad A = \frac{1}{2m}\vec{\pi}^2 . \tag{31}$$

Obviously, these operators are identical up to an additional constant $M_\pm = A + mc^2$ and hence the condition (9); that is, $[M_\pm, A] = 0$ is trivially fulfilled. In other words, the Klein–Gordon Hamiltonian (30) for a charged scalar particle in the presence of an arbitrary magnetic field represents a supersymmetric relativistic spin-zero Hamiltonian. Let us note that the two operators (31) in essence are given by the Landau Hamiltonian $H_{\mathrm{L}} := (\vec{p} - e\vec{A}/c)^2/2m = \vec{\pi}^2/2m$ of a non-relativistic spinless charged particle of mass m in a magnetic field; that is, $M_\pm = H_{\mathrm{L}} + mc^2$ and $A = H_{\mathrm{L}}$. As a consequence, the eigenvalue problem for (30) is reduced to that of H_{L}. The FW transformed Hamiltonian explicitly reads

$$H_{\mathrm{FW}} = \begin{pmatrix} \sqrt{(H_{\mathrm{L}} + mc^2)^2 - H_{\mathrm{L}}^2} & 0 \\ 0 & -\sqrt{(H_{\mathrm{L}} + mc^2)^2 - H_{\mathrm{L}}^2} \end{pmatrix} = \beta mc^2 \sqrt{1 + \frac{2H_{\mathrm{L}}}{mc^2}} . \tag{32}$$

Let us denote the eigenvalues of H_{L} by ϵ, then the eigenvalues of $M_\pm$ and $H_\pm = H_{\mathrm{L}}^2/2mc^2$ read

$$m_\pm c^2 = mc^2 + \epsilon, \qquad \varepsilon = \frac{\epsilon^2}{2mc^2} \tag{33}$$

and via relation (27) we find the eigenvalues of the Klein–Gordon Hamilonian

$$E_\pm = \pm\sqrt{m_\pm^2 c^4 - 2mc^2\varepsilon} = \pm\sqrt{(mc^2+\epsilon)^2 - \epsilon^2} = \pm mc^2\sqrt{1 + \frac{2\epsilon}{mc^2}}, \tag{34}$$

which is a result expected from relation (32). For a non-vanishing constant magnetic field, say in the z-direction $\vec{B} = B\vec{e}_z$ with $B \neq 0$, the eigenvalues of H_{L} are the well-known Landau levels [34,35]

$$\epsilon = \hbar\omega_c\left(n + \frac{1}{2}\right) + \frac{\hbar^2 k_z^2}{2m}, \qquad n \in \mathbb{N}_0, \qquad k_z \in \mathbb{R}, \qquad \omega_c := \frac{|eB|}{mc}. \tag{35}$$

As $\epsilon \geq \hbar\omega_c/2 > 0$ so is $\varepsilon > 0$ and hence $\dim\ker H_- = \dim\ker H_- = 0$. In other words, the Witten index (23) vanishes and SUSY is broken for the Klein–Gordon Hamiltonian in a constant magnetic field.

4.2. The Dirac Hamiltonian with Magnetic Field

The Dirac equation representing the relativistic dynamics of spin-$\frac{1}{2}$ fermions has intensively been studied since its introduction. See, for example, the excellent book by Thaller [29]. For a charged particle in an arbitrary external magnetic field, the Dirac Hamiltonian reads

$$H = \begin{pmatrix} mc^2 & c\vec{\sigma}\cdot\vec{\pi} \\ c\vec{\sigma}\cdot\vec{\pi} & -mc^2 \end{pmatrix}, \tag{36}$$

where $\vec{\sigma} = (\sigma_1, \sigma_2, \sigma_3)^T$ stands for a three-dimensional vector who's components are given by the Pauli matrices acting on $\mathbb{C}^2$, thus representing the spin-$\frac{1}{2}$ degree of freedom. Comparing this with the general form (8), we may identify the operators $M_\pm = mc^2$ and $A = c\sigma\cdot\vec{\pi} = A^\dagger$ and note that condition (9) is trivially fulfilled. Hence, the Dirac Hamiltonian (36) is indeed supersymmetric and its FW transformed form is known [32] to be expressible in terms of the non-relativistic Pauli Hamiltonian for a spin-$\frac{1}{2}$ particle with Landé g-factor $g = 2$.

$$H_{\mathrm{P}} := \frac{1}{2m}(\vec{\sigma}\cdot\vec{\pi})^2 = \frac{1}{2m}(\vec{p} - e\vec{A}/c)^2 - \frac{e\hbar}{mc}\vec{B}\cdot\vec{\sigma}. \tag{37}$$

Obviously, the partner Hamiltonians can be identified with the Pauli Hamiltonian $H_\pm = A^2/2mc^2 = H_{\mathrm{P}}$ and we find for the FW transformed Dirac Hamiltonian

$$H_{\mathrm{FW}} = \begin{pmatrix} \sqrt{m^2 c^4 + 2mc^2 H_{\mathrm{P}}} & 0 \\ 0 & -\sqrt{m^2 c^4 + 2mc^2 H_{\mathrm{P}}} \end{pmatrix} = \beta mc^2\sqrt{1 + \frac{2H_{\mathrm{P}}}{mc^2}}. \tag{38}$$

As shown by Aharonov and Casher [36], for a magnetic field having only a z-component, $\vec{B} = B(x,y)\vec{e}_z$, the ground-state energy of H_{P} is zero and the Witten index (23) in essence is given by the flux $F = \int_{\mathbb{R}^2} dxdy B(x,y)$ measured in units of the magnetic flux quantum $\Phi_0 := 2\pi\hbar c/|e|$. Hence, SUSY is unbroken in this case. For details, see for example [31], where it is shown that the Pauli Hamiltonian for this magnetic field provides an additional SUSY structure within both subspaces $\mathcal{H}_\pm$.

Finally, for an electron ($e < 0$) in a constant magnetic field, the eigenvalues of H_{P} are well-known

$$\epsilon = \hbar\omega_c\left(n + \frac{1}{2} + s_z\right) + \frac{\hbar^2 k_z^2}{2m}, \qquad n \in \mathbb{N}_0, \qquad k_z \in \mathbb{R}, \qquad s_z \in \left\{-\tfrac{1}{2}, \tfrac{1}{2}\right\}, \tag{39}$$

and the eigenvalues for (36) determined via (27) are the relativistic Landau levels first obtained by Rabi in 1928 [37]

$$E_{\pm} = \pm\sqrt{m^2c^2 + \hbar^2c^2k_z^2 + 2mc^2\hbar\omega_c(n + 1/2 + s_z)}.\tag{40}$$

The degeneracy for each set of quantum numbers (n, s_z, k_z) is given by the largest integer which is strictly less than $|F|/\Phi_0$ and is only finite in case the magnetic field has a compact support.

4.3. The Spin-1 Hamiltonian with Magnetic Field

Initiated by Proca's work [5], several authors have studied relativistic spin-one wave equations. Let us mention here the work by Duffin [14], by Kemmer [15] and by Yukawa, Sakata and Taketani [16,17]. An early study of the eigenvalue problem is due to Corben and Schwinger [18]. A relativistic Hamiltonian was studied, for example, by Young and Bludman [38], by Krase et al. [39] and by Tsai and coworkers [40,41]. The later work by Daicic and Frankel [42] presents an alternative solution to eigenvalue problem of the spin-one Hamiltonian in an external magnetic field. A recent treatment via FW transformation can be found in ref. [43].

The Hamiltonian of a charged spin-one particle with a priori arbitrary g-factor is given by, see for example [38,42,43],

$$H = \begin{pmatrix} mc^2 + \frac{\vec{\pi}^2}{2m} - \frac{ge\hbar}{2mc}(\vec{S}\cdot\vec{B}) & \frac{\vec{\pi}^2}{2m} - \frac{1}{m}(\vec{S}\cdot\vec{\pi})^2 + \frac{(g-2)e\hbar}{2mc}(\vec{S}\cdot\vec{B}) \\ -\frac{\vec{\pi}^2}{2m} + \frac{1}{m}(\vec{S}\cdot\vec{\pi})^2 - \frac{(g-2)e\hbar}{2mc}(\vec{S}\cdot\vec{B}) & -mc^2 - \frac{\vec{\pi}^2}{2m} + \frac{ge\hbar}{2mc}(\vec{S}\cdot\vec{B}) \end{pmatrix},\tag{41}$$

where $\vec{S} = (S_1, S_2, S_3)^T$ is a vector who's components are 3×3 matrices acting on $\mathbb{C}^3$ and obeying the $SO(3)$ algebra $[S_i, S_j] = i\varepsilon_{ijk}S_k$ representing the spin-one-degree of freedom of the particle. Again, we may identify the operators

$$M_{\pm} := mc^2 + \frac{\vec{\pi}^2}{2m} - \frac{ge\hbar}{2mc}(\vec{S}\cdot\vec{B}), \qquad A := \frac{\vec{\pi}^2}{2m} - \frac{1}{m}(\vec{S}\cdot\vec{\pi})^2 + \frac{(g-2)e\hbar}{2mc}(\vec{S}\cdot\vec{B}) = A^{\dagger}.\tag{42}$$

From now on, let us assume that the magnetic field $\vec{B}$ is constant, i.e., $\vec{A} = \frac{1}{2}\vec{B}\times\vec{r}$. Under this condition, one may verify that

$$[M_{\pm}, A] = (g-2)\frac{e\hbar}{2m^2c}\left[(\vec{S}\cdot\vec{B}), (\vec{S}\cdot\vec{\pi})^2\right].\tag{43}$$

Hence, the SUSY condition (9) is fulfilled if and only if $g = 2$. In other words, the relativistic spin-one Hamiltonian (41) is a supersymmetric Hamiltonian if the gyromagnetic factor is given by $g = 2$. For a detailed discussion, we refer to the recent paper [44]. Here, we remark that the "Vector Boson" Hamiltonian

$$H_V := \frac{\vec{\pi}^2}{2m} - \frac{e\hbar}{mc}(\vec{S}\cdot\vec{B})\tag{44}$$

represents the non-relativistic Hamiltonian of a charged spin-one particle in a magnetic field with gyromagnetic factor $g = 2$. Note that (44) is related to the quantity α introduced by Weaver [45] by $H_V = \frac{\alpha}{2m}$. That this is indeed the non-relativistic version of (41) was already mentioned in ref. [39]. With this we have $M_{\pm} = H_V + mc^2$ and with relation (3.18) and (3.19) from ref. [42], see also the Appendix A, a straightforward calculation shows that $A^2 = H_V^2$. That is, the partner Hamiltonians read $H_{\pm} = H_V^2/2mc^2$ and the transformed FW Hamiltonian takes the form

$$H_{FW} = \begin{pmatrix} \sqrt{(H_V + mc^2)^2 - H_V^2} & 0 \\ 0 & -\sqrt{(H_V + mc^2)^2 - H_V^2} \end{pmatrix} = \beta mc^2\sqrt{1 + \frac{2H_V}{mc^2}}\tag{45}$$

The eigenvalues of (44) are given by (we assume $e < 0$)

$$\epsilon = \hbar\omega_c\left(n + \frac{1}{2} + s_z\right) + \frac{\hbar^2 k_z^2}{2m}, \qquad n \in \mathbb{N}_0, \qquad k_z \in \mathbb{R}, \qquad s_z \in \{-1,0,1\}. \tag{46}$$

Hence, the spectrum of the partner Hamiltonians $H_\pm$ is given by $\varepsilon = \epsilon^2/2mc^2$ and SUSY is unbroken as $\epsilon = 0$ for $n = 0$, $s_z = -1$ and $k_z = \pm 1/\lambda_L$ with $\lambda_L := \sqrt{\hbar/m\omega_c} = \sqrt{\hbar c/|eB|}$ being the Lamor length [42]. That is, SUSY is unbroken for a spin-1 particle in a homogeneous magnetic field but the Witten index remains zero as $H_+ = H_-$ and therefore $\dim\ker H_+ = \dim\ker H_-$. The corresponding eigenvalues of (41) are given by

$$E_\pm = \pm\sqrt{m^2c^2 + \hbar^2 c^2 k_z^2 + 2mc^2\hbar\omega_c(n + 1/2 + s_z)}, \tag{47}$$

which is identical in form to the Dirac case (40) but s_z now taking the integer values as given in (46). In fact, for $k_z = 0$, $n = 0$ and $s_z = -1$, the above eigenvalues would become complex if $|B| > m^2c^3/|e|\hbar$. Such large magnetic fields would imply $\lambda_L < \lambda_C := \hbar/mc$; that is, the Lamor wavelength being smaller than the Compton wavelength of the vector boson. Note that confining a quantum particle to a region of the order of its Compton wavelength $\Delta x \sim \lambda_C$ implies by the uncertainty relation a momentum fluctuation $\Delta p \sim mc$ and thus a single particle description is no longer appropriate. In other words for such large magnetic fields a description via quantum field theory must be applied.

5. The Resolvent of Supersymmetric Relativistic Arbitrary-Spin Hamiltonians

In this section, we want to study the resolvent or Green's function of supersymmetric relativistic arbitrary-spin Hamiltonians defined as

$$G(z) := \frac{1}{H - z}, \qquad z \in \mathbb{C}\backslash\mathrm{spec}\,H. \tag{48}$$

For this, is it convenient to first look at the iterated resolvent which is given by

$$g(\zeta) := \frac{1}{H^2 - \zeta}, \qquad \zeta \in \mathbb{C}\backslash\mathrm{spec}\,H^2 \tag{49}$$

and is related with (48) via the obvious relation

$$G(z) = (H + z)\,g(z^2). \tag{50}$$

H^2 is block-diagonal and so is g; hence, it can be put into the form

$$g(\zeta) := \begin{pmatrix} g^+(\zeta) & 0 \\ 0 & g^-(\zeta) \end{pmatrix} \qquad \text{with} \qquad g^\pm(\zeta) := \frac{1}{M_\pm^2 + (-1)^{2s+1}2mc^2 H_\pm - \zeta}. \tag{51}$$

As a result, the resolvent (48) can be expressed in terms of (51) as follows:

$$G(z) = \begin{pmatrix} (z + M_+)g^+(z^2) & Ag^-(z^2) \\ (-1)^{2s+1}A^\dagger g^+(z^2) & (z - M_-)g^-(z^2) \end{pmatrix}. \tag{52}$$

In the following subsections, we will explicitly consider the three cases discussed in the previous section. It will turn out that for these three cases, the diagonal elements $g^\pm$ of the iterated Green's function can be expressed in terms of the Green's function of the corresponding non-relativistic Hamiltonian H_{NR}; that is,

$$g^{\pm}(\zeta) = \frac{1}{2mc^2} G_{NR}\left(\frac{\zeta}{2mc^2} - \frac{mc^2}{2}\right), \qquad G_{NR}(\xi) := \frac{1}{H_{NR} - \xi}, \qquad \xi \in \mathbb{C} \backslash \operatorname{spec} H_{NR}. \tag{53}$$

Note that the relation $\xi = z^2/2mc^2 - mc^2/2$, which can be put into the form $z = \pm mc^2\sqrt{1 + 2\xi/mc^2}$, in essence reflects the relation (28).

5.1. The Resolvent of the Klein–Gordon Hamiltonian with Magnetic Field

Following the discussion of the first example of Section 4, we may express all relevant operators in terms of the Landau Hamiltonian $H_L = \vec{\pi}^2/2m$. Explicitly, we have

$$M_{\pm} = H_L + mc^2, \qquad H_{\pm} = H_L^2/2mc^2, \qquad A = H_L\dagger, \tag{54}$$

which results in the iterated resolvents

$$g^{\pm}(\zeta) := \frac{1}{(H_L + mc^2)^2 - H_L^2 - \zeta} = \frac{1}{2mc^2} G_L\left(\frac{\zeta}{2mc^2} - \frac{mc^2}{2}\right), \tag{55}$$

where G_L stands for the Green function of the Landau Hamiltonian in terms of which the Klein–Gordon Hamiltonian reads

$$H = \begin{pmatrix} mc^2 + H_L & H_L \\ -H_L & -mc^2 - H_L \end{pmatrix}. \tag{56}$$

The Green's function then reads in terms of the Landau Hamiltonian

$$G(z) = \frac{1}{2mc^2}\begin{pmatrix} z + mc^2 + H_L & H_L \\ -H_L & z - mc^2 - H_L \end{pmatrix} G_L\left(\frac{z^2}{2mc^2} - \frac{mc^2}{2}\right). \tag{57}$$

5.2. The Resolvent of the Dirac Particle in a Magnetic Field

As in the above discussion, let us first recall the observations made in Section 4.2; that is,

$$M_{\pm} = mc^2, \qquad H_{\pm} = A^2/2mc^2 = H_P, \qquad A = c\vec{\sigma}\cdot\vec{\pi}, \tag{58}$$

which provide us with the components of the iterated kernel

$$g^{\pm}(\zeta) := \frac{1}{2mc^2 H_P + m^2c^4 - \zeta} = \frac{1}{2mc^2} G_P\left(\frac{\zeta}{2mc^2} - \frac{mc^2}{2}\right), \tag{59}$$

where $G_P(\epsilon) := (H_P - \epsilon)^{-1}$ is the resolvent of the non-relativistic Pauli Hamiltonian. In terms of this Pauli Green's function and the spin projection operator A, the Dirac Green's function can be put into the form

$$G(z) = \frac{1}{2mc^2}\begin{pmatrix} z + mc^2 & A \\ A & z - mc^2 \end{pmatrix} G_P\left(\frac{z^2}{2mc^2} - \frac{mc^2}{2}\right). \tag{60}$$

Some explicit examples have been worked out in ref. [32].

5.3. The Resolvent of a Vector Boson in a Magnetic Field

From Section 4.3, let us recall the relevant operators as follows:

$$M_{\pm} = H_V + mc^2, \qquad H_{\pm} = H_V^2/2mc^2, \qquad A = \frac{\vec{\pi}^2}{2m} - \frac{1}{m}(\vec{S}\cdot\vec{\pi})^2, \tag{61}$$

where the vector Hamilton H_V is given in Equation (44). Recalling that $A^2 = H_V^2$, we find for the iterated Green's functions

$$g^{\pm}(\zeta) := \frac{1}{(H_V + mc^2)^2 - H_V^2 - \zeta} = \frac{1}{2mc^2} G_V\left(\frac{\zeta}{2mc^2} - \frac{mc^2}{2}\right) \tag{62}$$

with $G_V(\epsilon) := (H_V - \epsilon)^{-1}$. The relativistic spin-one Hamiltonian explicitly reads

$$H = \begin{pmatrix} mc^2 + H_V & A \\ -A & -mc^2 - H_V \end{pmatrix} \tag{63}$$

and leads us to the Green's function

$$G(z) = \frac{1}{2mc^2} \begin{pmatrix} z + mc^2 + H_V & A \\ -A & z - mc^2 - H_V \end{pmatrix} G_V\left(\frac{z^2}{2mc^2} - \frac{mc^2}{2}\right). \tag{64}$$

6. Summary and Outlook

In this work we have considered relativistic one-particle Hamiltonians for an arbitrary but fixed spin s and have shown that under the condition, that its even part commutes with its odd part, a SUSY structure can be established. Here, the SUSY transformations map states of negative energy to those of positive energy and vice versa. This is different to the usual SUSY concepts in quantum field theory where those charges transform bosonic into fermionic states and vice versa. As examples, we have chosen the physically most relevant cases of a massive charged particle in a magnetic field for the cases of a scalar particle ($s = 0$), a Dirac fermion ($s = 1/2$) and a vector boson ($s = 1$). In the case of a constant magnetic field, SUSY is broken for $s = 0$ but remains unbroken for $s = 1/2$ and $s = 1$. The Witten index is only non-zero in the Dirac case but vanishes for the bosonic cases discussed. However, all three cases have resulted in the notable observation (28) that the FW-transformed Hamiltonian H_{FW} is entirely expressible in terms of a corresponding non-relativistic Hamiltonian H_{NR}. As $H_{\mathrm{FW}}^2 = H^2$, the relativistic energy–momentum relation can be put into the form

$$H^2 = m^2 c^4 + 2mc^2 H_{\mathrm{NR}}, \tag{65}$$

which allows us to relate H_{NR} with the SUSY Hamiltonian (11).

There naturally arises the desire to also study the higher-spin cases $s \geq 3/2$. The corresponding free-particle Hamiltonians have been constructed, for example, by Guertin [46] in a unified way. However, as Guertin mentions, only for the cases discussed here, i.e., $s = 0, 1/2$ and 1, the corresponding Hamiltonians are local operators.

Another route for further investigation would be to consider more exotic magnetic fields. For example, choosing an imaginary vector potential such that the kinetic momentum takes the form $\vec{\pi} = \vec{p} + im\omega\vec{r}$ in essence leads for $s = 1/2$ to the so-called Dirac oscillator, which is know to exhibit such a SUSY structure [32]. To the best of our knowledge, the corresponding Klein–Gordon and vector boson oscillators have not yet been studied in the context of SUSY. Similarly, following the discussion of ref. [32] on the Dirac case, one may extend these discussion on a path-integral representation of the iterated Green's functions to the bosonic cases $s = 0$ and $s = 1$.

Funding: This research received no external funding.

Acknowledgments: The author has enjoyed enlightening discussions with Simone Warzel for which I am very thankful. The comments made by the referees are very much appreciated.

Conflicts of Interest: The author declares no conflicts of interest.

Appendix A. Some Useful Relations for the Spin-One Case

In this Appendix A we present a few relations which provide some additional steps used in Section 4.3. For an arbitrary magnetic field let us recall that the components of the kinetic momentum given by $\pi_j = p_j - (e/c)A_j$ obey the commutation relation

$$[\pi_k, \pi_l] = (i\hbar e/c)\,\varepsilon_{klm}B_m \tag{A1}$$

where we use Einstein's summation convention for repeated indices. From this relation one may derive the commutator $[\pi_k, \vec{S}\cdot\vec{\pi}] = (i\hbar e/c)\,\varepsilon_{klm}S_l B_m$ which in turn leads us to

$$\left[\vec{\pi}^2, \vec{S}\cdot\vec{\pi}\right] = (e\hbar/c)[\vec{S}\cdot\vec{B}, \vec{S}\cdot\vec{\pi}] + (e\hbar/c)S_k S_l\,(\pi_l B_k - B_l \pi_k)\,. \tag{A2}$$

For an arbitrary magnetic field the components of the kinetic momentum do not commute with the components of the magnetic field. However, if we now assume that the magnetic field is constant one may commute in the last term these components. That is, under the assumption that $\vec{B} = const.$ we arrive at

$$\left[\vec{\pi}^2, \vec{S}\cdot\vec{\pi}\right] = (2e\hbar/c)[\vec{S}\cdot\vec{B}, \vec{S}\cdot\vec{\pi}]\,, \tag{A3}$$

which in turn results in

$$\left[\vec{\pi}^2, (\vec{S}\cdot\vec{\pi})^2\right] = \frac{2e\hbar}{c}\left[(\vec{S}\cdot\vec{B}), (\vec{S}\cdot\vec{\pi})^2\right]\,. \tag{A4}$$

Note that relation (A3) was already given in Equation (3.17) of ref. [42]. With the help of (A4) it is easy to calculate the commutator

$$[M_\pm, A] = (g - 2)\frac{e\hbar}{2m^2 c}\left[(\vec{S}\cdot\vec{B}), (\vec{S}\cdot\vec{\pi})^2\right] + (2g - 2)\frac{e\hbar}{2m^2 c}\left[\vec{\pi}^2, (\vec{S}\cdot\vec{B})\right]\,. \tag{A5}$$

Noting that we have derived this under the assumption of a constant magnetic field the last commutator in above expression vanishes and hence we arrive at Equation (43). Note that $[S_k, S_l] = i\varepsilon_{klm}S_m$ and therefore the first term on the right-hand-side above even for a constant magnetic field only vanishes when $g = 2$.

With the assumption that the magnetic field is constant and utilising below properties of the spin-one matrices

$$S_i S_j S_k + S_k S_j S_i = \delta_{ij}S_k + \delta_{jk}S_i\,, \qquad \varepsilon_{ijk}S_i S_j B_k = i\vec{S}\cdot\vec{B} \tag{A6}$$

one may verify the relations (see Equations (3.18) and (3.19) in ref. [42])

$$\begin{aligned}
(\vec{S}\cdot\vec{\pi})^4 &= \left(\vec{\pi}^2 - \tfrac{2e\hbar}{c}(\vec{S}\cdot\vec{B})\right)(\vec{S}\cdot\vec{\pi})^2 + \tfrac{e\hbar}{c}(\vec{B}\cdot\vec{\pi})(\vec{S}\cdot\vec{\pi})\,,\\
\left\{(\vec{S}\cdot\vec{B}), (\vec{S}\cdot\vec{\pi})^2\right\} &= \left(\vec{\pi}^2 - \tfrac{e\hbar}{c}(\vec{S}\cdot\vec{B})\right)(\vec{S}\cdot\vec{B}) + (\vec{B}\cdot\vec{\pi})(\vec{S}\cdot\vec{\pi})\,.
\end{aligned} \tag{A7}$$

Noting that for $g = 2$ we have

$$H_V := \frac{\vec{\pi}^2}{2m} - \frac{e\hbar}{mc}(\vec{S}\cdot\vec{B})\,, \qquad A = \frac{\vec{\pi}^2}{2m} - \frac{1}{m}(\vec{S}\cdot\vec{\pi})^2 \tag{A8}$$

and with above relations (A7) immediately follows that $A^2 = H_V^2$ as claimed in the main text. Finally, let us mention the explicit form of the energy eigenfunctions can also be found in ref. [42].

References

1. Klein, O. Quantentheorie und fünfdimensionale Relativitätstheorie. *Z. Phys.* **1926**, *37*, 895–906.
2. Gordon, W. Der Comptoneffekt nach der Schrödingerschen Theorie. *Z. Phys.* **1926**, *40*, 117–133.
3. Dirac, P.A.M. The Quantum Theory of the Electron. *Proc. R. Soc. A* **1928**, *117*, 610–624.

4.   Dirac, P.A.M. The Quantum Theory of the Electron Part II. *Proc. R. Soc. A* **1928**, *118*, 351–361.

5.   Proca, A. Sur la théorie ondulatoire des électrons positifs et négatifs. *J. Phys. Radium* **1936**, *7*, 347–353.

6.   Dirac, P.A.M. Relativistic Wave Equations. *Proc. R. Soc. A* **1936**, *155*, 447–459.

7.   Fierz, M. Über die relativistische Theorie kräftefreier Teilchen mit beliebigem Spin. *Helv. Phys. Acta* **1939**, *12*, 3–37.

8.   Fierz, M.; Pauli, W. On Relativistic Wave Equations for Particles of Arbitrary Spin in an Electromagnetic Field. *Proc. R. Soc. A* **1939**, *173*, 211–232.

9.   Bhabha, H.J. Relativistic wave equations for the elementary particles. *Rev. Mod. Phys.* **1945**, *17*, 200–216.

10. Bargmann, V.; Wigner, E.P. Group Theoretical Discussion of Relativistic Wave Equations. *Proc. Natl. Acad. Sci. USA* **1948**, *34*, 211–223.

11. Foldy, L.L.; Wouthuysen, S.A. On the Dirac Theory of Spin 1/2 Particles and Its Non-Relativistic Limit. *Phys. Rev.* **1950**, *78*, 29–36.

12. Foldy, L.L. Synthesis of Covariant Particle Equations. *Phys. Rev.* **1956**, *102*, 568–581.

13. Feshbach, H.; Villars, F. Elementary Relativistic Wave Mechanics of Spin 0 and Spin 1/2 Particles. *Rev. Mod. Phys.* **1958**, *30*, 24–45.

14. Duffin, R.J. On The Characteristic Matrices of Covariant Systems. *Phys. Rev.* **1938**, *54*, 1114.

15. Kemmer, N. The particle aspect of meson theory. *Proc. R. Soc. A* **1939**, *173*, 91–116.

16. Yukawa, H.; Sakata, S.; Taketani, M. On the Interaction of Elementary Particles. III. *Proc. Phys.-Math. Soc. Jpn.* **1938**, *20*, 319–340.

17. Sakata, S.; Taketani, M. On the Wave Equation of Meson. *Proc. Phys.-Math. Soc. Jpn.* **1940**, *22*, 757–770.

18. Corben, H.C.; Schwinger, J. The Electromagnetic Properties of Mesotrons. *Phys. Rev.* **1940**, *58*, 953–968.

19. Schrödinger, E. The wave equation for spin 1 in Hamiltonian form. *Proc. R. Soc. A* **1955**, *229*, 39–43.

20. Silenko, A. Exact form of the exponential Foldy-Wouthuysen transformation operator for an arbitrary-spin particle. *Phys. Rev. A* **2016**, *94*, 032104.

21. Simulik, V.M. Relativistic Equations for arbitray Spin, especially for the Spin $s = 2$. *Ukr. J. Phys.* **2019**, *64*, 1064–1068.

22. Nicolai, H. Supersymmetry and spin systems. *J. Phys. A* **1976**, *9* 1497–1506.

23. Witten, E. Dynamical breaking of supersymmetry. *Nucl. Phys. B* **1981**, *188* 513–554.

24. Cooper, F.; Khare, A.; Sukhatme, U. Supersymmetry and quantum mechanics. *Phys. Rep.* **1995**, *251*, 267–385.

25. Junker, G. *Supersymmetric Methods in Quantum and Statistical Physics*, 1st ed.; Springer: Berlin, Germany, 1996.

26. Jackiw, R. Fractional charge and zero modes for planar systems in a magnetic field. *Phys. Rev. D* **1984**, *29* 2375–2377.

27. Ui, H. Supersymmetric quantum mechanics and fermion in a gauge field of $(1 + 2)$ dimensions. *Prog. Theor. Phys.* **1984**, *72*, 192–193.

28. Sarma, S.D.; Adam, S.; Hwang, E.H.; Rossi, E. Electronic transport in two-dimensional graphene. *Rev. Mod. Phys.* **2011**, *83* 407–470.

29. Thaller, B. *The Dirac Equation*; Springer: Berlin, Germany, 1992.

30. Znoil, M. Relativistic supersymmetric quantum mechanics based on Klein-Gordon equation. *J. Phys. A* **2004**, *37*, 9557.

31. Junker, G. *Supersymmetric Methods in Quantum, Statistical and Solid State Physics, Enlarged and Revised Edition*; IOP Publishing: Bristol, UK, 2019.

32. Junker, G.; Inomata, A. Path integral and spectral representations for supersymmetric Dirac-Hamiltonians *J. Math. Phys.* **2018**, *59*, 052301.

33. Witten, E. Constraints on supersymmetry breaking. *Nucl. Phys. B* **1982**, *202* 253–316.

34. Fock, V. Bemerkung zur Quantelung des harmonischen Oszillators im Magnetfeld. *Z. Phys.* **1928**, *47* 446–448.

35. Landau, L. Diamagnetismus der Metalle. *Z. Phys.* **1930**, *64*, 629–637.

36. Aharonov, Y.; Casher, A. Ground state of a spin-$\frac{1}{2}$ charged particle in a two-dimensional magnetic field. *Phys. Rev. A* **1978**, *19*, 2461–2462.

37. Rabi, I.I. Das freie Elektron im homogenen Magnetfeld nach der Diracschen Theorie. *Z. Phys.* **1928**, *49*, 507–511.

38. Young, J.A.; Bludman, S.A. Electromagnetic Properties of a Charged Vector Meson. *Phys. Rev.* **1963**, *131*, 2326–2334.

39. Krase, L.D.; Lu, P.; Good, R.H., Jr. Stationary States of a Spin-1 Particle in a Constant Magnetic Field. *Phys. Rev. D* **1971**, *3*, 1275–12798.

40. Tsai, W.-Y.; Yildiz, A. Motion of Charged Particles in a Homogeneous Magnetic Field. *Phys. Rev. D* **1971** *4*, 3643–3648.

41. Goldman, T.; Tsai, W.-Y. Motion of Charged Particles in a Homogeneous Magnetic Field II. *Phys. Rev. D* **1971**, *4*, 3648–3651.

42. Daicic, J.; Frankel, N.E. Relativistic spin-1 bosons in a magnetic field. *J. Phys. A Math. Gen.* **1993**, *26*, 1397–1408.

43. Silenko, A.J. High precision description and new properties of a spin-1 particle in a magnetic field. *Phys. Rev. D* **2014**, *89*, 121701.

44. Junker, G. Supersymmetric quantum mechanics requires $g = 2$ for vector bosons. *Eur. Phys. Lett.* **2020**, *130*, 30003.

45. Weaver, D.L. Application of a single-particle-theory calculation of $g - 2$ to spin one. *Phys. Rev. D* **1976**, *14*, 2824–2825.

46. Guertin, R.F. Relativistic Hamiltonian Equations for Any Spin. *Ann. Phys.* **1974**, *88*, 504–553.

 *symmetry*

Article

On the Supersymmetry of the Klein–Gordon Oscillator [†]

Georg Junker [1,2]

1   European Southern Observatory, Karl-Schwarzschild-Straße 2, D-85748 Garching, Germany; gjunker@eso.org
2   Institut für Theoretische Physik I, Universität Erlangen-Nürnberg, Staudtstraße 7,
     D-91058 Erlangen, Germany
†   Dedicated to Akira Inomata on the Occasion of His 90th Birthday.

Abstract: The three-dimensional Klein–Gordon oscillator exhibits an algebraic structure known from supersymmetric quantum mechanics. The supersymmetry is unbroken with a vanishing Witten index, and it is utilized to derive the spectral properties of the Klein–Gordon oscillator, which is closely related to that of the nonrelativistic harmonic oscillator in three dimensions. Supersymmetry also enables us to derive a closed-form expression for the energy-dependent Green's function.

Keywords: Klein–Gordon oscillator; supersymmetric quantum mechanics; Green's function

Citation: Junker, G. On the Supersymmetry of the Klein–Gordon Oscillator. *Symmetry* **2021**, *13*, 835. https://doi.org/10.3390/sym13050835

Academic Editor: Themis Bowcock

Received: 14 April 2021
Accepted: 7 May 2021
Published: 10 May 2021

Publisher's Note: MDPI stays neutral with regard to jurisdictional claims in published maps and institutional affiliations.

1. Introduction

Starting with Galileo's pendulum experiment [1] in 1602, and with Hook's law of elasticity [2] from 1678, harmonic oscillators played significant roles in classical physics. More importantly, the harmonic oscillator was the first system to which early quantum theory was successfully applied by Planck [3] in 1900 when developing his law of black body radiation. Nowadays, the harmonic oscillator is a standard part of any introductory text book on nonrelativistic quantum mechanics. In relativistic quantum mechanics, the harmonic oscillator was initially studied within Dirac's theory of electrons in the 1960s [4–6], but attracted considerable attention only with the seminal work by Moshinsky and Szczepaniak [7] (see also Quesne and Moshinsky [8]). Inspired by this so-called Dirac oscillator, the Klein–Gordon oscillator (KGO) was studied by various authors [9–11].

The KGO Hamiltonian characterises a relativistic spin-zero particle with mass m minimally coupled to a complex linear vector potential. Since its introduction, the KGO has attracted much interest. The spectral properties of the one-dimensional system were discussed, for example, in [12,13]. For a treatment in noncommutative space, see [14,15]; for recent results in a nontrivial topology, see [16–19] and the references therein.

Since 1990, the Dirac oscillator has been known to exhibit a supersymmetric (SUSY) structure that, in turn, allows for explicit solutions [20–23]. More recently, SUSY also enabled us to formulate Feynman's path integral approach for Dirac systems [24]. SUSY in the current context is not based on the original idea, which transforms between states with different internal spin-degree of freedom, but refers to what is commonly known nowadays as supersymmetric quantum mechanics; see, for example, [25] and the references therein.

The purpose of the present work is twofold. First, we show that the Klein–Gordon oscillator possesses a hidden SUSY in the aforementioned sense. Second, we derive an explicit expression for the Green function of the KGO. In doing so, we closely follow the generic approach for SUSY in relativistic Hamiltonians with fixed but arbitrary spin [26].

In the next section, we set up the stage with a brief discussion on the KGO Hamiltonian in three space dimensions and show that this Hamiltonian exhibits a SUSY structure by mapping it onto a quantum mechanical SUSY system. This is then utilized to derive explicit results of the system. In Section 3, we derive the eigenvalues and associated eigenstates. In Section 4, we derive the corresponding Green's function in a closed form. Lastly, Section 5 closes with a summary and some comments.

2. Supersymmetry

The Hamiltonian form of the Klein–Gordon equation with arbitrary vector potential was originally introduced by Feshbach and Villars [27], from which the KGO Hamiltonian may be constructed via minimal coupling $\vec{p} \rightarrow \vec{\pi} := \vec{p} - \mathrm{i}m\omega\vec{r}$, where $m > 0$ stands for the mass of the spinless Klein–Gordon particle, and $\omega > 0$ is a coupling constant to be identified with the harmonic oscillator frequency. This minimal coupling might be interpreted as a complex-valued vector potential of form $\vec{A}(\vec{r}) := \mathrm{i}(mc\omega/q)\vec{r}$, with q being the particle charge, and c the speed of light. However, such a vector potential is not linked to any kind of gauge invariance, as $\vec{A}(\vec{r}) = \mathrm{i}(mc\omega/2q)\vec{\nabla}r^2$ cannot be gauged away by a pure phase factor in the wave function due to the presence of the imaginary unit.

First explicit expressions of the KGO Hamiltonian were presented by Debergh et al. [9] for an isotropic system. The KGO Hamiltonian for a more general anisotropic oscillator system is due to Bruce and Minning [10]. For the sake of simplicity, we consider the isotropic system characterised by Hamiltonian

$$\mathcal{H} := \frac{\vec{\pi}^{\dagger} \cdot \vec{\pi}}{2m} \otimes (\tau_3 + \mathrm{i}\tau_2) + mc^2 \otimes \tau_3 \tag{1}$$

acting on Hilbert space $L^2(\mathbb{R}^3) \otimes \mathbb{C}^2$. In the above, τ_is stand for Pauli matrices

$$\tau_1 := \begin{pmatrix} 0 & 1 \\ 1 & 0 \end{pmatrix}, \qquad \mathrm{i}\tau_2 := \begin{pmatrix} 0 & 1 \\ -1 & 0 \end{pmatrix}, \qquad \tau_3 := \begin{pmatrix} 1 & 0 \\ 0 & -1 \end{pmatrix}. \tag{2}$$

These Pauli matrices do not represent a spin degree of freedom. The 2-spinors on which the above Hamiltonian acts are those originally introduced by Feshbach and Villars [27].

The KGO Hamiltonian (1) is pseudo-Hermitian [28,29], that is, $\mathcal{H}^{\dagger} = \tau_3\,\mathcal{H}\,\tau_3$, and reads in an explicit 2×2 matrix notation

$$\mathcal{H} = \begin{pmatrix} M & A \\ -A & -M \end{pmatrix}, \tag{3}$$

where we set $M := H_{\mathrm{NR}} + mc^2$ and $A := H_{\mathrm{NR}}$, both in essence being represented by the nonrelativistic harmonic oscillator Hamiltonian in three dimensions

$$H_{\mathrm{NR}} := \frac{\vec{\pi}^{\dagger} \cdot \vec{\pi}}{2m} = \frac{1}{2m}\vec{p}^{\,2} + \frac{m}{2}\omega^2\vec{r}^{\,2} - \frac{3}{2}\hbar\omega. \tag{4}$$

Here and in the following, we use calligraphic symbols for operators acting on the full Hilbert space $L^2(\mathbb{R}^3) \otimes \mathbb{C}^2$, and operators represented in italics act on subspace $L^2(\mathbb{R}^3)$.

Obviously, diagonal and off-diagonal elements in (3) commute, i.e., $[M, A] = 0$. Hence, following the general approach of [26], an $N = 2$ SUSY structure can be established as follows.

$$\mathcal{H}_{\mathrm{SUSY}} := \frac{1}{2mc^2}\left(\mathcal{M}^2 - \mathcal{H}^2\right) = \frac{1}{2mc^2}H_{\mathrm{NR}}^2 \otimes 1,$$

$$\mathcal{Q} := \frac{1}{\sqrt{2mc^2}}\begin{pmatrix} 0 & A \\ 0 & 0 \end{pmatrix}, \qquad \mathcal{W} := \begin{pmatrix} 1 & 0 \\ 0 & -1 \end{pmatrix} \equiv \tau_3, \tag{5}$$

where we set $\mathcal{M} := M \otimes 1$. The above SUSY operators obey SUSY algebra

$$\mathcal{H}_{\mathrm{SUSY}} = \{\mathcal{Q}, \mathcal{Q}^{\dagger}\}, \qquad \mathcal{Q}^2 = 0 = \mathcal{Q}^{\dagger 2},$$

$$[\mathcal{W}, \mathcal{H}_{\mathrm{SUSY}}] = 0, \qquad \{\mathcal{Q}, \mathcal{W}\} = 0 = \{\mathcal{Q}^{\dagger}, \mathcal{W}\}. \tag{6}$$

In the current context, the third Pauli matrix plays the role of the Witten party operator $\mathcal{W}$. Therefore, the upper and lower components of a general 2-spinor belong to the subspace with positive and negative Witten parity, respectively. We further remark that dim ker $\mathcal{Q}$ = dim ker $\mathcal{Q}^\dagger$ = dim ker H_{NR} = 1. That is, SUSY is unbroken, as $\mathcal{H}_{\mathrm{SUSY}}$ has zero-energy eigenstates [25], but Witten index Δ still vanishes as

$$\Delta := \mathrm{ind}\,\mathcal{Q} = \dim \ker \mathcal{Q} - \dim \ker \mathcal{Q}^\dagger = 0. \tag{7}$$

To the best of our knowledge, this is the first quantum mechanical system with an unbroken $N = 2$ SUSY but vanishing Witten index, implying that the spectrum of $\mathcal{H}$ is fully symmetric with respect to the origin, as we see in the following section.

3. Spectral Properties

As was recently shown [26], SUSY in a relativistic Hamiltonian implies the existence of a Foldy–Wouthuysen transformation, which brings that Hamiltonian into a block-diagonal form. In the case of the KGO, this transformation operator $\mathcal{U}$, which is a pseudounitary operator in the sense that $\mathcal{U}^{-1} = \tau_3\,\mathcal{U}^\dagger\,\tau_3$, reads

$$\mathcal{U} := \frac{|\mathcal{H}| + \tau_3\mathcal{H}}{\sqrt{2(\mathcal{H}^2 + M|\mathcal{H}|)}} \tag{8}$$

leading to block-diagonal Foldy–Wouthuysen Hamiltonian

$$\mathcal{H}_{\mathrm{FW}} := \mathcal{U}\,\mathcal{H}\,\mathcal{U}^{-1} = H_{\mathrm{FW}} \otimes \tau_3\,, \qquad H_{\mathrm{FW}} := \sqrt{2mc^2 H_{\mathrm{NR}} + m^2 c^4}\,. \tag{9}$$

To be more explicit, let us define $\tanh\Theta := A/M = H_{\mathrm{NR}}/(H_{\mathrm{NR}} + mc^2)$, which then allows for us to write transformation (8) in matrix form [9]

$$\mathcal{U} = \begin{pmatrix} \cosh\frac{\Theta}{2} & \sinh\frac{\Theta}{2} \\ \sinh\frac{\Theta}{2} & \cosh\frac{\Theta}{2} \end{pmatrix} = \frac{1}{\sqrt{2}} \begin{pmatrix} \sqrt{\frac{M}{H_{\mathrm{FW}}} + 1} & \sqrt{\frac{M}{H_{\mathrm{FW}}} - 1} \\ \sqrt{\frac{M}{H_{\mathrm{FW}}} - 1} & \sqrt{\frac{M}{H_{\mathrm{FW}}} + 1} \end{pmatrix}. \tag{10}$$

The above expressions are functions of operators of which all may be expressed in terms of H_{NR}. Hence, using the spectral theorem, these are well-defined. In fact, with the spectral properties of the nonrelativistic harmonic-oscillator Hamiltonian (4), one can directly obtain those of (1). Let $\psi_{n\ell\mu}$ denote the well-known eigenfunctions of H_{NR} corresponding to eigenvalue $\varepsilon_{n\ell}$; then, we have

$$H_{\mathrm{NR}}\,\psi_{n\ell\mu} = \varepsilon_{n\ell}\,\psi_{n\ell\mu}\,, \qquad \varepsilon_{n\ell} = \hbar\omega(2n + \ell)\,, \qquad n, \ell \in \mathbb{N}_0\,,$$

$$\psi_{n\ell\mu}(\vec{r}) = \left(\frac{m\omega}{\hbar}\right)^{\ell/2+3/4} \sqrt{\frac{2\,n!}{\Gamma(n+\ell+3/2)}}\, r^\ell\, e^{-m\omega r^2/\hbar}\, L_n^{\ell+1/2}\!\left(\frac{m\omega}{\hbar}r^2\right) Y_{\ell\mu}(\vec{e})\,, \tag{11}$$

$$\mu \in \{-\ell, -\ell+1, \ldots, \ell-1, \ell\}\,, \qquad r := |\vec{r}|\,, \qquad \vec{e} := \vec{r}/r\,,$$

where $L_n^{\ell+1/2}$ and $Y_{\ell\mu}$ denote the associated Laguerre polynomials and spherical harmonics, respectively. See, for example, ref. [30]. The eigenvalues and eigenfunctions of (1) are explicitly given by

$$\mathcal{H}\Psi_{n\ell\mu}^\pm = E_{n\ell}^\pm\Psi_{n\ell\mu}^\pm\,, \qquad E_{n\ell}^\pm = \pm mc^2\sqrt{1 + \frac{2\varepsilon_{n\ell}}{mc^2}}\,,$$

$$\Psi_{n\ell\mu}^+(\vec{r}) = \psi_{n\ell\mu}(\vec{r})\begin{pmatrix} \cosh\frac{\vartheta_{n\ell}}{2} \\ -\sinh\frac{\vartheta_{n\ell}}{2} \end{pmatrix}, \qquad \Psi_{n\ell\mu}^-(\vec{r}) = \psi_{n\ell\mu}(\vec{r})\begin{pmatrix} -\sinh\frac{\vartheta_{n\ell}}{2} \\ \cosh\frac{\vartheta_{n\ell}}{2} \end{pmatrix}, \tag{12}$$

where $\tanh\vartheta_{n\ell} := \varepsilon_{n\ell}/(\varepsilon_{n\ell} + mc^2)$. These states form an orthonormal basis in $L^2(\mathbb{R}^2) \otimes \mathbb{C}^2$ with respect to the scalar product [27].

$$\langle\Psi_1|\Psi_2\rangle := \int_{\mathbb{R}^3} \mathrm{d}^3\vec{r}\,\overline{\Psi}_1(\vec{r})\,\tau_3\,\Psi_2(\vec{r})\,, \tag{13}$$

where the overbar stands for the transposed and complex conjugated 2-spinor. That is,

$$\langle \Psi^{\pm}_{n\ell\mu} | \Psi^{\pm}_{n'\ell'\mu'} \rangle = \pm \delta_{nn'} \delta_{\ell\ell'} \delta_{\mu\mu'}, \qquad \langle \Psi^{\pm}_{n\ell\mu} | \Psi^{\mp}_{n'\ell'\mu'} \rangle = 0. \tag{14}$$

Obviously, the scalar product (13), which was already introduced by Feshbach and Villars [27], is not positive definite and might raise some questions on its probabilistic interpretation. However, Mostafazadeh's theory of pseudo-Hermitian operators provides a solution for this obstacle. The Klein–Gordon case was explicitly discussed in [28,29].

The SUSY ground states associated with nonrelativistic eigenvalue $\varepsilon_{00} = 0$ are given by

$$\Psi^{+}_{000}(\vec{r}) = \left(\frac{m\omega}{\hbar\pi} \right)^{3/4} e^{-m\omega r^2/\hbar} \begin{pmatrix} 1 \\ 0 \end{pmatrix}, \qquad \Psi^{-}_{000}(\vec{r}) = \left(\frac{m\omega}{\hbar\pi} \right)^{3/4} e^{-m\omega r^2/\hbar} \begin{pmatrix} 0 \\ 1 \end{pmatrix} \tag{15}$$

with corresponding eigenvalues $E^{\pm}_{00} = \pm mc^2$. The Foldy–Wouthuysen Hamiltonian (9) can be written as

$$\mathcal{H}_{\mathrm{FW}} = mc^2 \sqrt{1 + \frac{2H_{\mathrm{NR}}}{mc^2}} \otimes \tau_3, \tag{16}$$

a form already observed for other relativistic Hamiltonians exhibiting a SUSY [26].

4. Green's Function

The SUSY established for the KGO in the previous section also allows for us to study Green's function associated with the KGO Hamiltonian (1). Following the general approach of [26], Green's function, defined by

$$\mathcal{G}(z) := \frac{1}{\mathcal{H} - z}, \qquad z \in \mathbb{C} \backslash \mathrm{spec}\, \mathcal{H}, \tag{17}$$

can be expressed in terms of iterated Green's function $\mathcal{G}_{\mathrm{I}}$, that is,

$$\mathcal{G}(z) = (\mathcal{H} + z)\mathcal{G}_{\mathrm{I}}(z^2), \qquad \mathcal{G}_{\mathrm{I}}(z^2) := \frac{1}{\mathcal{H}^2 - z^2}. \tag{18}$$

Noting that $\mathcal{H}^2 = 2mc^2(H_{\mathrm{NR}} + mc^2/2) \otimes 1$, the iterated Green's function can be written in terms of nonrelativistic Green's function $G_{\mathrm{NR}}(\varepsilon) := (H_{\mathrm{NR}} - \varepsilon)^{-1}$ associated with H_{NR}, as follows.

$$\mathcal{G}_{\mathrm{I}}(z^2) = \frac{1}{2mc^2} G_{\mathrm{NR}}(\varepsilon) \otimes 1, \qquad \varepsilon := \frac{z^2}{2mc^2} - \frac{mc^2}{2} = 2mc^2 \left(\left(\frac{z}{2mc^2} \right)^2 - \frac{1}{4} \right). \tag{19}$$

Inserting this into above relation (18) results in

$$\mathcal{G}(z) = \frac{1}{2mc^2} \begin{pmatrix} (H_{\mathrm{NR}} + mc^2 + z)\, G_{\mathrm{NR}}(\varepsilon) & H_{\mathrm{NR}}\, G_{\mathrm{NR}}(\varepsilon) \\ -H_{\mathrm{NR}}\, G_{\mathrm{NR}}(\varepsilon) & -(H_{\mathrm{NR}} + mc^2 - z)\, G_{\mathrm{NR}}(\varepsilon) \end{pmatrix}. \tag{20}$$

Using defining relation $H_{\mathrm{NR}}\, G_{\mathrm{NR}}(\varepsilon) = \varepsilon\, G_{\mathrm{NR}}(\varepsilon)$ with the second relation in (19) leads us to closed-form expression

$$\mathcal{G}(z) = G_{\mathrm{NR}}(\varepsilon) \begin{pmatrix} \left(\frac{1}{2} + \frac{z}{2mc^2} \right)\left(\frac{1}{2} + \frac{z}{2mc^2} \right) & \left(\frac{1}{2} + \frac{z}{2mc^2} \right)\left(\frac{z}{2mc^2} - \frac{1}{2} \right) \\ \left(\frac{1}{2} + \frac{z}{2mc^2} \right)\left(\frac{1}{2} - \frac{z}{2mc^2} \right) & \left(\frac{1}{2} - \frac{z}{2mc^2} \right)\left(\frac{z}{2mc^2} - \frac{1}{2} \right) \end{pmatrix}, \tag{21}$$

with ε as defined in (19). The reader is invited to verify that $\mathcal{H}\, \mathcal{G}(z) = z\, \mathcal{G}(z)$. With definition $\tanh \vartheta := \varepsilon/(\varepsilon + mc^2)$, the above result may be placed into form

$$\mathcal{G}(z) = \frac{G_{\mathrm{NR}}(\varepsilon)}{\cosh \frac{\vartheta}{2} - \sinh \frac{\vartheta}{2}} \begin{pmatrix} \cosh^2 \frac{\vartheta}{2} & \cosh \frac{\vartheta}{2} \sinh \frac{\vartheta}{2} \\ -\cosh \frac{\vartheta}{2} \sinh \frac{\vartheta}{2} & -\sinh^2 \frac{\vartheta}{2} \end{pmatrix}. \tag{22}$$

The coordinate representation of $G_{NR}(\vec{r}, \vec{r}', \varepsilon) := \langle \vec{r} | G_{NR}(\varepsilon) | \vec{r}' \rangle$ has been known for long (see, for example [31]) and explicitly reads

$$G_{NR}(\vec{r}, \vec{r}', \varepsilon) = \frac{1}{rr'} \sum_{\ell=0}^{\infty} G_\ell(r, r', \varepsilon) \sum_{\mu=-\ell}^{\ell} Y_{\ell\mu}^*(\vec{e}') Y_{\ell\mu}(\vec{e}),$$

$$G_\ell(r, r', \varepsilon) = -\frac{\Gamma\left(\frac{\ell}{2} - \frac{\varepsilon}{2\hbar\omega}\right)}{\sqrt{rr'}\,\hbar\omega} W_{\lambda,\nu}\left(r_>^2\, m\omega/\hbar\right) M_{\lambda,\nu}\left(r_<^2\, m\omega/\hbar\right),$$

(23)

where $W_{\lambda,\nu}$ and $M_{\lambda,\nu}$ denote Whittaker's functions, and we set $\lambda := \frac{\varepsilon}{2\hbar\omega} + \frac{3}{4}$, $\nu := \frac{\ell}{2} + \frac{1}{4}$, $r_> := \max\{r, r'\}$ and $r_> := \min\{r, r'\}$.

5. Summary and Outlook

In this work, we showed that the KGO exhibits a SUSY structure, closely following the general approach of [26]. The SUSY of the KGO was found to be unbroken but with a vanishing Witten index. Despite eigenvalues in (12) having been known for a long time (see, for example, [9,10]), the associated eigenstates in (12) have, to our knowledge, never been presented. In [10] only the eigenstates of $\mathcal{H}_{FW}$ were given. In addition, SUSY enabled us to calculate the KGO Green function in a closed form.

Obviously, the current discussion for an isotropic oscillator may be extended to that for the anisotropic oscillator following Bruce and Manning [10]. Here, in essence, one needs to reduce the problem to three one-dimensional harmonic oscillators. An explicit expression for Green's function may also be obtained, as the one-dimensional harmonic oscillator Green's function is also known in closed form. See, for example, Glasser and Nieto [32], and the discussion of the associated Dirac problem [33]. One may also pursue the path integral approach for the KGO along the lines of the corresponding approach for the Dirac oscillator [24]. Another route for further investigation would be to look at generalised nonharmonic oscillators characterized by a potential function $U_a(r) := \lambda_a r^a$ using minimal substitution $\vec{\pi} := \vec{p} - i(\vec{\nabla} U_a)(r)$. Such power-law potentials obey duality symmetry in classical and nonrelativistic quantum mechanics (see recent work [34] and references therein). In particular, a harmonic potential where $a = 2$, which corresponds to the discussed KGO case, is dual to the Kepler potential where $a = -1$.

Another extension is to apply the current SUSY construction to the relativistic $S = 1$ oscillator. However, as was argued by Debergh et al. [9], the diagonal and off-diagonal matrix elements of the associated Hamiltonian no longer commute. Hence, it may not be possible to establish a SUSY structure.

Funding: This research received no external funding.

Conflicts of Interest: The author declares no conflict of interest.

References

1. Palmieri, P. A phenomenology of Galileo's experiments with pendulums. *Br. J. Hist. Sci.* **2009**, *42*, 479–513. [CrossRef]
2. Hook, R. *De Potentia Restitutiva, or of Spring. Explaining the Power of Springing Bodies*, The Royal Society. London, UK, 1678.
3. Planck, M. Zur Theorie des Gesetzes der Energieverteilung im Normalspectrum. *Verhandl. Dtsch. Phys. Ges.* **1900**, *2*, 237–245.
4. Itô, D.; Mori, K.; Carriere, E. An example of dynamical systems with linear trajectory. *Nuovo Cim. A* **1967**, *51*, 1119–1121. [CrossRef]
5. Swamy, N.V.V.J. Exact Solution of the Dirac Equation with an Equivalent Oscillator Potential. *Phys. Rev.* **1969**, *180*, 1225–1226. [CrossRef]
6. Cook, P.A. Relativistic harmonic oscillators with intrinsic spin structure. *Lett. Nuovo Cim.* **1971**, *1*, 419–426. [CrossRef]
7. Moshinsky, M.; Szczepaniak, A. The Dirac oscillator. *J. Phys. A* **1989**, *22*, L817–L819. [CrossRef]
8. Quesne C.; Moshinsky, M. Symmetry Lie algebra of the Dirac oscillator. *J. Phys. A* **1990**, *23*, 2263–2272. [CrossRef]
9. Debergh, N.; Ndimubandi, J.; Strivay, D. On relativistic scalar and vector mesons with harmonic oscillatorlike interactions. *Z. Phys. C* **1992**, *56*, 421–425. [CrossRef]
10. Bruce S.; Minning, P. The Klein–Gordon Oscillator. *Nuovo Cim. A* **1993**, *106*, 711–713. [CrossRef]
11. Dvoeglazov, V.V.; del Sol Mesa, A. Notes on oscillator-like interactions of various spin relativistic particles. *NASA Conf. Pub.* **1994**, *3286*, 333–340.

12. Rao, N.A.; Kagali, B.A. Energy profile of the one-dimensional Klein–Gordon oscillator. *Phys. Scr.* **2008**, *77*, 015003. [CrossRef]
13. Boumali, A.; Hafdallah, A.; Toumi, A. Comment on 'Energy profile of the one-dimensional Klein–Gordon oscillator'. *Phys. Scr.* **2011**, *84*, 037001. [CrossRef]
14. Mirza, B.; Mohadesi, M.; Zare, S. The Klein–Gordon and the Dirac Oscillators in a Noncommutative Space. *Commun. Theor. Phys.* **2004**, *42*, 664–668. [CrossRef]
15. Mirza, B.; Mohadesi, M. Relativistic Oscillators in a Noncommutative Space and in a Magnetic Field. *Commun. Theor. Phys.* **2011**, *55*, 405–409. [CrossRef]
16. Santos, L.C.N.; Mota, C.E.; Barros, C.C., Jr. Klein–Gordon Oscillator in a Topologically Nontrivial Space-Time. *Adv. High Energy Phys.* **2019**, *2019*, 2729352. [CrossRef]
17. Ahmed, F. The generalized Klein–Gordon oscillator in the background of cosmic string space-time with a linear potential in the Kaluza–Klein theory. *Eur. Phys. J. C* **2020**, *80*, 211. [CrossRef]
18. Bragança, E.A.F.; Vitória, R.L.L.; Belich, H.; Bezerra de Mello, E.R. Relativistic quantum oscillators in the global monopole spacetime. *Eur. Phys. J.* **2020**, *80*, 206. [CrossRef]
19. Zhong, L.; Chen, H.; Hassanabadi, H.; Long, Z.-W.; Long, C.-Y. The study of the generalized Klein–Gordon oscillator in the context of the Som-Raychaudhuri space-time. *arXiv* **2021**, arXiv:2101.06356.
20. Benítez, J.; Martínez y Romero, R.P.; Núñez-Yépez, H.N.; Salas-Brito, A.L. Solution and Hidden Super-symmetry of a Dirac Oscillator. *Phys. Rev. Lett.* **1990**, *64*, 1643–1645; ERRATA *Phys. Rev. Lett.* **1990**, *65*, 2085. [CrossRef]
21. Beckers J.; Debergh, N. Supersymmetry, Foldy-Wouthuysen trausformations, and relativistic oscillators. *Phys. Rev. D* **1990**, *42*, 1255–1259. [CrossRef] [PubMed]
22. Martínez y Romero, R.P.; Morena, M.; Zentella, A. Supersymmetric properties and stability of the Dirac sea. *Phys. Rev. D* **1991**, *43*, 2036–2040. [CrossRef]
23. Quesne, C. Supersymmetry and the Dirac oscillator. *Int. J. Mod. Phys.* **1991**, *6*, 1567–1589. [CrossRef]
24. Junker, G.; Inomata, A. Path Integral and Spectral Representations for Supersymmetric Dirac-Hamiltonians. *J. Math. Phys.* **2018**, *59*, 052301. [CrossRef]
25. Junker, G. *Supersymmetric Methods in Quantum, Statistical and Solid State Physics*, Enlarged and revised ed.; IOP Publishing: Bristol, UK, 2019.
26. Junker, G. Supersymmetry of Relativistic Hamiltonians for Arbitrary Spin. *Symmetry* **2021**, *12*, 1590. [CrossRef]
27. Feshbach, H.; Villars, F. Elementary Relativistic Wave Mechanics of Spin 0 and Spin 1/2 Particles. *Rev. Mod. Phys.* **1958**, *30*, 24–45. [CrossRef]
28. Mostafazadeh, A. Is Pseudo-Hermitian Quantum Mechanics an Indefinite-Metric Quantum Theory? *Czech. J. Phys.* **2003**, *53*, 1079–1084. [CrossRef]
29. Mostafazadeh, A. A Physical Realization of the Generalized PT-, C-, and CPT-Symmetries and the Position Operator for Klein–Gordon Fields. *Int. J. Mod. Phys. A* **2006**, *53*, 2553–2572. [CrossRef]
30. Galindo, A.; Pascual, P. *Quantum Mechanics I*; Springer: Berlin, Germany, 1990.
31. Blinder, S.M. Propagators from integral representations of Green's functions for the N-dimensional free particle, harmonic oscillator and Coulomb problems. *J. Math. Phys.* **1984**, *25*, 905–909. [CrossRef]
32. Glasser, M.L.; Nieto, L.M. The energy level structure of a variety of one-dimensional confining potentials and the effects of a local singular perturbation. *Can. J. Phys.* **2015**, *93*, 1588–1596. [CrossRef]
33. Junker, G. Supersymmetric Dirac Hamiltonians in $(1+1)$ dimensions revisited. *Eur. Phys. J. Plus* **2020**, *135*, 464. [CrossRef]
34. Inomata, A.; Junker, G. Power Law Duality in Classical and Quantum Mechanics. *Symmetry* **2021**, *13*, 409. [CrossRef]

 *symmetry*

Article

Power Law Duality in Classical and Quantum Mechanics

Akira Inomata [1] and Georg Junker [2,3,*]

1   Department of Physics, State University of New York at Albany, Albany, NY 12222, USA;
     ainomata@albany.edu
2   European Southern Observatory, Karl-Schwarzschild-Straße 2, D-85748 Garching, Germany
3   Institut für Theoretische Physik I, Universität Erlangen-Nürnberg, Staudtstraße 7,
     D-91058 Erlangen, Germany
*   Correspondence: gjunker@eso.org

Abstract: The Newton–Hooke duality and its generalization to arbitrary power laws in classical, semiclassical and quantum mechanics are discussed. We pursue a view that the power-law duality is a symmetry of the action under a set of duality operations. The power dual symmetry is defined by invariance and reciprocity of the action in the form of Hamilton's characteristic function. We find that the power-law duality is basically a classical notion and breaks down at the level of angular quantization. We propose an ad hoc procedure to preserve the dual symmetry in quantum mechanics. The energy-coupling exchange maps required as part of the duality operations that take one system to another lead to an energy formula that relates the new energy to the old energy. The transformation property of the Green function satisfying the radial Schrödinger equation yields a formula that relates the new Green function to the old one. The energy spectrum of the linear motion in a fractional power potential is semiclassically evaluated. We find a way to show the Coulomb–Hooke duality in the supersymmetric semiclassical action. We also study the confinement potential problem with the help of the dual structure of a two-term power potential.

Keywords: power-law duality; classical and quantum mechanics; semiclassical quantization; supersymmetric quantum mechanics; quark confinement

check for
updates

Citation: Inomata, A.; Junker, G. Power Law Duality in Classical and Quantum Mechanics. *Symmetry* **2021**, *13*, 409. https://doi.org/10.3390/sym13030409

Academic Editor: Georg Junker

Received: 24 January 2021
Accepted: 26 February 2021
Published: 3 March 2021

1. Introduction

In recent years, numerous exoplanets have been discovered. One of the best Doppler spectrographs to discover low-mass exoplanets using the radial velocity method are HARPS (High Accuracy Radial Velocity Planet Searcher) installed on ESO's 3.6 m telescope at La Silla and ESPRESSO (Echelle Spectrograph for Rocky Exoplanet- and Stable Spectroscopic Observations) installed on ESO's VLT at Paranal Observatory in Chile. See, e.g., [1,2]. NASA's Kepler space telescope has discovered more than half of the currently known exoplanets using the so-called transit method. See, e.g., [3,4]. For some theoretical work on planetary systems see, e.g., [5]. In exoplanetary research it is a generally accepted view that Newton's law of gravitation holds in extrasolar systems [6]. Orbit mechanics of exoplanets, as is the case of solar planets and satellites, is classical mechanics of the Kepler problem under small perturbations. The common procedure for the study of perturbations to the Kepler motion is the so-called regularization, introduced by Levi–Civita (1906) for the planar motion [7,8] and generalized by Kustaanheimo and Stiefel (1965) to the spatial motion [9]. The regularization in celestial mechanics is a transformation of the singular equation of motion for the Kepler problem to the non-singular equation of motion for the harmonic oscillator problem with or without perturbations. It identifies the Kepler motion with the harmonic oscillation, assuring the dual relation between Newton's law and Hooke's law (here, following the tradition, we mean by Newton's law the inverse-square force law of gravitation and by Hooke's law the linear force law for the harmonic oscillation. Although Hooke found the inverse square force law for gravitation prior to Newton, he was short of skills in proving that the orbit of a planet is an ellipse in accordance with Kepler's first law, while Newton was able not only to confirm that the inverse square force law yields an elliptic orbit but also to show conversely that the inverse square force law follows Kepler's first law. History gave Newton the full credit of the inverse square force law for gravitation. For a detailed account, see, e.g., Arnold's book [10]). The Newton–Hooke duality has been discussed by many authors from various aspects [11,12]. The basic elements of regularization are: (i) a transformation of space variables, (ii) interpretation of the conserved energy as the coupling constant, and (iii) a transformation of time parameter. The choice of space variables and time parameter is by no means unique. The transformation of space variables has been represented in terms of parabolic coordinates [7,8], complex numbers [13,14], spinors [9], quaternions [6,15,16], etc. The time transformation used by Sundman [13,17] and by Bohlin [14] (for Bohlin's theorem see also reference [10]) is essentially based on Newton's finding [18] that the areal speed dA/dt is constant for any central force motion. It takes the form $ds = Crdt$ where s is a fictitious time related to the eccentric anomaly. To improve numerical integrations for the orbital motion, a family of time transformations $ds = C_\eta r^\eta dt$, called generalized Sundman transformations, has also been discussed [19], in which s corresponds to the mean anomaly if $\eta = 0$, the eccentric anomaly if $\eta = 1$, the true anomaly if $\eta = 2$, and intermediate anomalies [20] for other values of η. Even more generalizing, a transformation of the form $ds = Q(r)dt$ has been introduced in the context of regularization [21].

As has been pointed out in the literature [10,18,22–24], the dual relation between the Kepler problem and the harmonic oscillator was already known in the time of Newton and Hooke. What Newton posed in their Principia was more general. According to Chandrasekhar's reading [18] out of the propositions and corollaries (particularly Proposition VII, Corollary III) in the Principia, Newton established the duality between the centripetal forces of the form, r^α and r^β, for the pairs $(\alpha, \beta) = (1, -2), (-1, -1)$ and $(-5, -5)$. Revisiting the question on the duality between a pair of arbitrary power forces, Kasner [25] and independently Arnol'd [10] obtained the condition, $(\alpha + 3)(\beta + 3) = 4$, for a dual pair. There are a number of articles on the duality of arbitrary power force laws [26,27]. Now on, for the sake of brevity, we shall refer to the duality of general power force laws as the power duality. The power duality includes the Newton–Hooke duality as a special case.

The quantum mechanical counterpart of the Kepler problem is the hydrogen atom problem. In 1926, Schrödinger [28,29] solved their equation for the hydrogen atom and successively for the harmonic oscillator. Although it must have been known that both radial equations for the hydrogen atom and for the harmonic oscillation are reducible to confluent hypergeometric equations [30], there was probably no particular urge to relate the Coulomb problem to the Hooke problem, before the interest in the accidental degeneracies arose [31–33]. Fock [31,32] pointed out that for the bound states the hydrogen atom has a hidden symmetry $SO(4)$ and an appropriate representation of the group can account for the degeneracy. In connection with Fock's work, Jauch and Hill [34] showed that the $2 - D$ harmonic oscillator has an algebraic structure of $su(2)$ which is doubly-isomorphic to the $so(3)$ algebra possessed by the $2 - D$ hydrogen atom. The transformation of the radial equation from the hydrogen atom to that of the harmonic oscillator or vice verse was studied by Schrödinger [35] and others, see Johnson's article [36] and references therein. The same problem in arbitrary dimensions has also been discussed from the supersymmetric interest [37]. In the post-Kustaanheimo–Stiefel (KS) era, the relation between the three dimensional Coulomb problem and the four dimensional harmonic oscillator was also investigated by implementing the KS transformation or its variations in the Schrödinger equation. See ref. [38] and references therein. The duality of radial equations with multi-terms of power potentials was studied in connection with the quark confinement [36,39,40].

The time transformation of the form $ds = C_\eta r^\eta dt$ used in classical mechanics is in principle integrable only along a classical trajectory. In other words, the fictitious time s is globally meaningful only when the form of $r(t)$ as a function of t is known. In quantum mechanics, such a transformation is no longer applicable due to the lack of classical paths. Hence it is futile to use any kind of time transformation formally to the time-dependent Schrödinger equation. The Schrödinger equation subject to the duality transformation is a time-independent radial equation possessing a fixed energy and a fixed angular momentum. The classical time transformation is replaced in quantum mechanics by a renormalization of the time-independent state function [41]. In summary, the duality transformation applicable to the Schrödinger equation consists of (i) a change of radial variable, (ii) an exchange of energy and coupling constant, and (iii) a transformation of state function. Having said so, when it comes to Feynman's path integral approach, we should recognize that the classical procedure of regularization prevails.

Feynman's path integral is based on the *c*-number Lagrangian and, as Feynman asserted [42], the path of a quantum particle for a short time d*t* can be regarded as a classical path. Therefore, the local time transformation associated with the duality transformation in classical mechanics can be revived in path integration. In fact, the Newton–Hooke duality plays an important role in path integration. Feynman's path integral in the standard form [42,43] provides a way to evaluate the transition probability from a point to another in space (the propagator or the Feynman kernel). The path integral in the original formulation gives exact solutions only for quadratic systems including the harmonic oscillator, but fails in solving the hydrogen atom problem. However, use of the KS transformation enables to convert the path integral for the hydrogen atom problem to that of the harmonic oscillator if the action of Feynman's path integral is slightly modified with a fixed energy term. In 1979, Duru and Kleinert [44], formally applying the KS transformation to the Hamiltonian path integral, succeeded to obtain the energy-dependent Green function for the hydrogen atom in the momentum representation. Again, with the help of the KS transformation, Ho and Inomata (1982) [45] carried out detailed calculations of Feynman's path integral with a modified action to derive the energy Green function in the coordinate representation. In 1984, on the basis of the polar coordinate formulation of path integral (1969) [46], without using the KS variables, the radial path integral for the hydrogen atom was transformed to that for the radial harmonic oscillator by Inomata for three dimensions [47] and by Steiner for arbitrary dimensions [48,49]. Since then a large number of examples have been solved by path integration [50,51]. Applications of the Newton–Hooke duality in path integration include those to the Coulomb problem on uniformly curved spaces [52,53], Kaluza–Klein monopole [54], and many others [51]. The idea of classical regularization also helped to open a way to look at the path integral from group theory and harmonic analysis [50,55,56]. The only work that discusses a confinement potential in the context of path integrals is Steiner's [57].

As has been briefly reviewed above, the Newton–Hooke duality and its generalizations have been extensively and exhaustively explored. In the present paper we pursue the dual relation (power-duality) between two systems with arbitrary power-law potentials from the symmetry point of view. While most of the previous works deal with equations of motion, we focus our attention on the symmetry of action integrals under a set of duality operations. Our duality discussion covers the classical, semiclassical and quantum-mechanical cases. In Section 2, we define the dual symmetry by invariance and reciprocity of the classical action in the form of Hamilton's characteristic function and specify a set of duality operations. Then we survey comprehensively the properties of the power-duality. The energy-coupling exchange relations contained as a part of the duality operations lead to various energy formulas. In Section 3, we bring the power-duality defined for the classical action to the semiclassical action for quantum mechanical systems. We argue that the power-duality is basically a classical notion and breaks down at the level of angular quantization. To preserve the basic idea of the dual symmetry in quantum mechanics, we propose as an ad hoc procedure to treat angular momentum L as a continuous parameter and to quantize it only after the transformation is completed. A linear motion in a fractional power-law potential is solved as an example to find the energy spectrum by extended use of the classical energy formulas. We also discussed the dual symmetry of the supersymmetric

(SUSY) semiclassical action. Although we are unable to verify general power duality, we find a way to show the Coulomb–Hooke symmetry in the SUSY semiclassical action. Section 4 analyzes the dual symmetry in quantum mechanics on the basis of an action having wave functions as variables. The energy formulas, eigenfunctions and Green functions for dual systems are discussed in detail, including the Coulomb–Hooke problem. We also explore a quark confinement problem as an application of multi-power potentials, showing that the zero-energy bound state in the confinement potential is in the power-dual relation with a radial harmonic oscillator. Section 5 gives a summary of the present paper and an outlook for the future work. Appendix A presents the Newton–Hooke–Morse triality that relates the Newton–Hooke duality to the Morse oscillator.

2. Power-Law Duality as a Symmetry

Duality is an interesting and important notion in mathematics and physics, but it has many faces [58]. In physics it may mean equivalence, complementarity, conjugation, correspondence, reciprocity, symmetry and so on. Newton's law and Hooke's law may be said dual to each other in the sense that a given orbit of one system can be mapped into an orbit of the other (one-to-one correspondence), whereas they may be a dual pair because the equation of motion of one system can be transformed into the equation of motion for the other (equivalence).

In this section, we pursue a view that the power duality is a symmetry of the classical action in the form of Hamiltonian's characteristic function, and discuss the power duality in classical, semiclassical and quantum mechanical cases.

2.1. Stipulations

Let us begin by proposing an operational definition of the power duality. We consider two distinct systems, A and B. System A (or A in short), characterized by an index or a set of indices a, consists of a power potential $V_a(r) \sim r^a$ and a particle of mass m_a moving in the potential with fixed angular momentum L_a and energy E_a. Similarly, system B (B in short), characterized by an index or a set of indices b, consists of a power potential $V_b(r) \sim r^b$ and a particle of mass m_b moving in the potential with fixed angular momentum L_b and energy E_b.

If there is a set of invertible transformations $\Delta(B, A)$ that takes A to B, then we say that A and B are equivalent. Naturally, the inverse of $\Delta(B, A)$ denoted by $\Delta(A, B) = \Delta^{-1}(B, A)$ takes B to A.

Let $X(a, b)$ and $X(b, a) = X^{-1}(a, b)$ be symbols for replacing the indices b by a and a by b, respectively. If B becomes A under $X(a, b)$ and A becomes B under $X(b, a)$, then we say that A and B are reciprocal to each other with respect to $\Delta(B, A)$. If A and B are equivalent and reciprocal, we say they are dual to each other. Since each of the two systems has a power potential, we regard the duality so stipulated as the power duality.

The successive applications of $\Delta(A, B)$ and $X(a, b)$ transform A to B and change B back to A. Consequently the combined actions leave A unchanged. In this sense we can view that the set of operations, $\{\Delta(A, B), X(a, b)\}$, or its inverse, $\{\Delta(B, A), X(b, a)\}$, is a symmetry operation for the power duality.

If a quantity Q_a belonging to system A transforms to Q_b while $\Delta(B, A)$ takes system A to system B, then we write $Q_b = \Delta(B, A)Q_a$. If Q_b can be converted to Q_a by $X(a, b)$, then we write $Q_a = X(a, b)Q_b$ and say that Q_a is form-invariant under $\Delta(B, A)$. If $Q_a = Q_b$, then Q_a is an invariant under $\Delta(B, A)$. If every Q_a belonging to system A is an invariant under $\Delta(B, A)$, then $\Delta(B, A)$ is an identity operation.

2.2. Duality in the Classical Action

The power duality in classical mechanics may be most easily demonstrated by considering the action integral of the form of Hamilton's characteristic function, $W(E) = S(t) + Et$, where S is the Hamilton's principal function and E is the energy of the system in question. The action is usually given by Hamilton's principal function,

$$S(\tau) = \int^{\tau} dt\, \mathcal{L} = \int^{\tau} dt \left[\frac{m}{2}\dot{\vec{r}}^2 - V(\vec{r})\right] \tag{1}$$

which leads to the Euler–Lagrange equations via Hamilton's variational principle. If the system is spherically symmetric, that is, if the potential $V(\vec{r})$ is independent of angular variables, then the action remains invariant under rotations. If the system is conservative, that is, if the Lagrangian is not an explicit function of time, then the action is invariant under time translations. In general, if the action is invariant under a transformation, then the transformation is often called a symmetry transformation.

For a conserved system, we can choose as the action Hamilton's characteristic function,

$$W(E) = \int^{\tau} dt\, \{\mathcal{L} + E\} = S(\tau) + E\tau, \qquad E = -\frac{\partial S(\tau)}{\partial \tau}. \tag{2}$$

Insofar as the system is conservative, both the principal action $S(\tau)$ and the characteristic action $W(E)$ yield the same equations of motion. For the radial motion of a particle of mass m with a chosen value of energy E and a chosen value of angular momentum L in a spherically symmetric potential $V(r)$, the radial action has the form,

$$W_{(r,t)}(E) = \int_{I_t} dt \left\{\frac{m}{2}\left(\frac{dr}{dt}\right)^2 - \frac{L^2}{2mr^2} - V(r) + E\right\}, \tag{3}$$

where $I_t = \tau(E) \ni t$ is the range of t. We let a system with a specific potential V_a be system A and append the subscript a to every parameter involved. In a similar manner, we let a system with V_b be system B whose parameters are all marked with a subscript b. For system A with a radial potential $V_a(r)$, we rewrite the action (3) in the form,

$$W_{(r,t)}(E_a) = \int_{I_\varphi} d\varphi \left(\frac{dt}{d\varphi}\right)\left\{\frac{m_a}{2}\left(\frac{dt}{d\varphi}\right)^{-2}\left(\frac{dr}{d\varphi}\right)^2 - \frac{L_a^2}{2m_a r^2} - U_a(r)\right\}, \tag{4}$$

with

$$U_a(r) = V_a(r) - E_a, \tag{5}$$

where φ is some fiducial time and $I_\varphi \ni \varphi$ is the range of integration.

In (4), as is often seen in the literature [36,40,41], we change the radial variable from r to ρ by a bijective differentiable map,

$$\mathfrak{R}_f: \quad r = f(\rho) \quad \Leftrightarrow \quad \rho = f^{-1}(r), \tag{6}$$

where f is a positive differentiable function of ρ, $0 < r < \infty$ and $0 < \rho < \infty$. With this change of variable we associate a change of time derivative from $(dt/d\varphi)$ to $(ds/d\varphi)$ by a bijective differentiable map,

$$\mathfrak{T}_g: \quad (dt/d\varphi) = g(\rho)(ds/d\varphi) \quad \Leftrightarrow \quad (ds/d\varphi) = \frac{(dt/d\varphi)}{g(f^{-1}(r))}. \tag{7}$$

In the above, we assume that both r and ρ are of the same dimension and that s has the dimension of time as t does. As a result of operations $\mathfrak{R}_f$ and $\mathfrak{T}_g$ on the action (4), we obtain

$$W_{(r,t)}(E_a) = \int_{I_\varphi} d\varphi \left(\frac{ds}{d\varphi}\right)\left\{\frac{m_a}{2}\frac{f'^2}{g}\left(\frac{ds}{d\varphi}\right)^{-2}\left(\frac{d\rho}{d\varphi}\right)^2 - \frac{gL_a^2}{2m_a f^2} - gU_a(f(\rho))\right\}, \tag{8}$$

whose implication is obscure till the transformation functions f and g are appropriately specified.

Suppose there is a set of operations Δ, including $\mathfrak{R}_f$ and $\mathfrak{T}_g$ as a subset, that can convert $W_{(r,t)}(E_a)$ of (8) to the form,

$$W_{(\rho,s)}(E_b) = \int_{I_\varphi} d\varphi \left(\frac{ds}{d\varphi} \right) \left\{ \frac{m_b}{2} \left(\frac{ds}{d\varphi} \right)^{-2} \left(\frac{d\rho}{d\varphi} \right)^2 - \frac{L_b^2}{2 m_b \rho^2} - U_b(\rho) \right\}, \tag{9}$$

with

$$U_b = V_b(\rho) - E_b, \tag{10}$$

where $V_b(\rho)$ is a real function of ρ, and E_b is a constant having the dimension of energy. Then we identify the new action (9) with the action of system B representing a particle of mass m_b which moves in a potential $V_b(\rho)$ with fixed values of angular momentum L_b and energy E_b. If $W_{\zeta_a}(E_a) = X(a,b) W_{\zeta_b}(E_b)$ where $\zeta_a = (r,t)$ and $\zeta_b = (\rho, s)$, then $W_{\zeta_a}(E_a)$ is form-invariant under Δ. Since $W_{(\rho,s)}(E_a)$ is physically identical with $W_{(r,t)}(E_a)$, if $W_{(\rho,s)}(E_a) = X(a,b) W_{(\rho,s)}(E_b)$, then we say that system A represented by $W_{(r,t)}(E_a)$ is dual to system B represented by $W_{(\rho,s)}(E_b)$ with respect to Δ.

2.3. Duality Transformations

In an effort to find such a set of operations Δ, we wish, as the first step, to determine the transformation functions $f(\rho)$ of (6) and $g(\rho)$ of (7) by demanding that the set of space and time transformations $\{\mathfrak{R}_f, \mathfrak{T}_g\}$ preserves the form-invariance of each term of the action. In other words, we determine $f(\rho)$ and $g(\rho)$ so as to retain (i) form-invariance of the kinetic term, (ii) form-invariance of the angular momentum term and (iii) form-invariance of the shifted potential term.

In the action $W_{(r,t)}(E_a)$ of (8), the functions $f(\rho)$ and $g(\rho)$ are arbitrary and independent of each other. To meet the condition (i), it is necessary that $g = \mu f'^2$ where μ is a positive constant. Then the kinetic term expressed in terms of the new variable can be interpreted as the kinetic energy of a particle with mass

$$\mathfrak{M}: \qquad m_b = m_a / \mu. \tag{11}$$

In order for the angular momentum term to keep its inverse square form as required by (ii), the transformation functions are to be chosen as

$$f(\rho) = C_\eta \rho^\eta, \qquad g(\rho) = \mu C_\eta^2 \eta^2 \rho^{2\eta - 2}, \tag{12}$$

where η is a non-zero real constant and C_η is an η dependent positive constant which has the dimension of $r^{1-\eta}$ as r and ρ have been assumed to possess the same dimension. With (12), the angular momentum term of (8) takes the form, $L_b^2 / (2 m_b \rho^2)$, when the mass changes by $\mathfrak{M}$ of (11), and the angular momentum L_a transforms to

$$\mathfrak{L}: \qquad L_b = \eta L_a. \tag{13}$$

To date, the forms of $f(\rho)$ and $g(\rho)$ in (12) have been determined by the asserted conditions (i) and (ii), even before the potential is specified. This means that (iii) is a condition to select a potential $V(r)$ pertinent to the given form of $g(\rho)$. More explicitly, (iii) demands that $gU_a(r)$ must be of the form,

$$gU_a = V_b(\rho) - E_b, \tag{14}$$

where $V_b(\rho)$ is such that $V_a(\rho) = X(a,b) V_b(\rho)$. Therefore, the space-time transformation $\{\mathfrak{R}_f, \mathfrak{T}_g\}$ subject to the form-invariance conditions (i)–(iii) is only applicable to a system with a limited class of potentials.

The simplest potential that belongs to this class is the single-term power potential $V_a(r) = \lambda_a r^a$ where $\lambda_a \in \mathbb{R}$ and $a \in \mathbb{R}$. The corresponding shifted potential is given by

$$U_a(r) = \lambda_a r^a - E_a \tag{15}$$

which transforms with (12) into

$$gU_a(r) = \mu\lambda_a C_\eta^{a+2}\eta^2 \rho^{a\eta+2\eta-2} - \mu C_\eta^2 \eta^2 \rho^{2\eta-2} E_a. \tag{16}$$

Under the condition (iii) the expected form of the shifted potential is

$$U_b(\rho) = gU_a(r) = \lambda_b \rho^b - E_b, \tag{17}$$

where $\lambda_b \in \mathbb{R}$ and $b \in \mathbb{R}$. Comparison of (16) and (17) gives us only two possible combinations for the new exponents and the new coupling and energy,

$$b = a\eta + 2\eta - 2 \quad \text{and} \quad 2\eta - 2 = 0, \tag{18}$$

$$\lambda_b = \mu C_\eta^{a+2}\eta^2\lambda_a \quad \text{and} \quad E_b = \mu C_\eta^2\eta^2 E_a \tag{19}$$

and

$$b = 2\eta - 2 \quad \text{and} \quad a\,\eta + 2\eta - 2 = 0, \quad (a \neq -2), \tag{20}$$

$$\lambda_b = -\mu C_\eta^2\eta^2\rho^{2\eta-2}E_a \quad \text{and} \quad E_b = -\mu C_\eta^{a+2}\eta^2\lambda_a, \tag{21}$$

Note that $a = -2$ is included in the first combination but excluded from the second combination.

In the following, we shall examine the two possible combinations in more detail by expressing the admissible transformations in terms of the exponents,

$$\eta_1 = 1, \quad \eta_a = 2/(a+2) \quad (a \neq 0, -2), \tag{22}$$

and separating the set of η_a into two as

$$\eta_+ = \{\eta_a | a > -2\}, \quad \eta_- = \{\eta_a | a < -2\}. \tag{23}$$

Chandrasekhar in their book [18] represents a pair of dual forces by $(a - 1, b - 1)$. In a way analogous to their notation, we also use the notation (a, b) via η for a pair of the exponents of power potentials when system A and system B are related by a transformation with η. We shall put the subscript F to differentiate the pairs of dual forces from those of dual potentials as $(a - 1, b - 1)_F = (a, b)$ whenever needed. Caution must be exercised in interpreting $(0, 0)$ which may mean $\lim_{\varepsilon \to 0}(\pm\varepsilon, \pm\varepsilon)$, $\lim_{\varepsilon \to 0}(\pm\varepsilon, \mp\varepsilon)$ and purely $(0, 0)$ (see the comments in below Subsections). We shall refer to the sets of pairs (a, b) related to the first combination (18)–(19) and the second combination (20)–(21) as Class I and Class II, respectively.

2.3.1. Class I

Class I is the supplementary set of self-dual pairs. Equation (18) of the first combination implies

$$\mathfrak{C}_1 : \quad \eta_1 = 1, \quad a = b \in \mathbb{R}, \tag{24}$$

which is denoted by (a, a) via η_1. In this case, (12) yields $f(\rho) = C_1\rho$ and $g(\rho) = \mu C_1^2$ where C_1 and μ are arbitrary dimensionless constants. With these transformation functions, (6) and (7) lead to a set of space and time transformations whose scale factors depend on neither space nor time,

$$\mathfrak{R}_1 : \quad r = C_1\rho, \tag{25}$$

and

$$\mathfrak{T}_1: \quad (dt/d\varphi) = \mu C_1^2 (ds/d\varphi). \tag{26}$$

Associated with the space and time transformations (25) and (26) are the scale changes in coupling and energy, as shown by (19),

$$\mathfrak{E}_1: \quad \lambda_a \to \lambda_b = (\mu C_1^{a+2})\lambda_a, \qquad E_a \to E_b = (\mu C_1^2)E_a. \tag{27}$$

According to (11), the mass also changes its scale,

$$\mathfrak{M}_1: \quad m_b = m_a/\mu. \tag{28}$$

From (13) and (24) follows the scale-invariant angular momentum (we use the subscript 0 for trivial transformations representing an identity),

$$\mathfrak{L}_0: \quad L_b = L_a. \tag{29}$$

In this manner we obtain a set of operations $\Delta_1 = \{\mathfrak{E}_1, \mathfrak{R}_1, \mathfrak{T}_1, \mathfrak{E}_1, \mathfrak{M}_1, \mathfrak{L}_0\}$ that leaves form-invariant the action for the power potential system. System B reached from system A by Δ_1 can go back to system A by $X(a,b)$. Hence, system A is dual to system B. Notice, however, that Δ_1 leads to a self-dual pair (a,a) via η_1 for any given $a \in \mathbb{R}$. In particular, $(0,0) = \lim_{\varepsilon \to 0}(\pm\varepsilon, \mp\varepsilon)$.

Remark 1. *Class I consists of self-dual pairs (a,a) via η_1 for all $a \in \mathbb{R}$. All pairs in this class are supplemental in the sense that they are not traditionally counted as dual pairs. Since Δ_1 is a qualified set of operations for preserving the form-invariance of the action, we include self-dual pairs of Class I in order to extend slightly the scope of the duality discussion.*

Remark 2. *The space transformation $\mathfrak{R}_1$ of (25) is a simple scaling of the radial variable as $C_1 > 0$. The scaling is valid for any chosen positive value of C_1. Hence it can be reduced, as desired, to the identity transformation $r = \rho$ by letting $C_1 = 1$. Those dual pairs linked by scaling may be regarded as trivial.*

Remark 3. *The scale transformation with $C_1 > 0$ induces the time scaling $\mathfrak{T}_1$ whereas the time has its own scaling behavior. The change in time (26) integrates to $t = C_1\mu s + \nu$ where ν is a constant of integration. The resulting time equation may be understood as consisting of a time translation $t = t' + \nu$, a scale change due to the space scaling $t' = C_1 s'$, and an intrinsic time scaling $s' = \mu s$. The time translation, under which the energy has been counted as conserved, is implicit in $\mathfrak{T}_1$. The scale factor μ of time scaling, independent of space scaling, can take any positive value. If $C_1 = 1$ and $\mu = 1$, then $\mathfrak{T}_1$ becomes the identity transformation of time, $(dt/d\varphi) = (ds/d\varphi)$.*

Remark 4. *The scale change in mass $m_b = \mu m_a$ is only caused by the intrinsic time scaling $t = \mu s$. If $\mu = 1$, then the mass of the system is conserved. Conversely, if $m_a = m_b$ is preferred, the time scaling with $\mu = 1$ must be chosen. The time scaling in classical mechanics has no particular significance. In fact, it adds nothing significant to the duality study. Therefore, in addition to the form-invariant requirements (i)–(iii), we demand (iv) the mass invariance $m_a = m_b = m$ by choosing $\mu = 1$. In this setting the time scaling occurs only in association with the space-scaling. In accordance with the condition (iv), we shall deal with systems of an invariant mass m for the rest of the present paper.*

Remark 5. *If $C_1 = 1$ and $\mu = 1$, then operations, $\mathfrak{E}_1$, $\mathfrak{M}_1$, and $\mathfrak{L}_0$, become identities of respective quantities. Thus, Δ_1 for $C_1 = 1$ and $\mu = 1$ is the set of identity operations, which we denote Δ_0. The set of operations Δ_1 for $C_1 > 0$ is trivial in the sense that it is reducible to the set of identity operations Δ_0.*

Remark 6. *If Class I is based only on the scale transformation, it may not be worth pursuing. As will be discussed in the proceeding sections, there are some examples that do not belong to the list of traditional dual pairs (Class II). In an effort to accommodate those exceptional pairs within the present scheme for the duality discussion, we look into the details hidden behind the space identity transformation $r = \rho$. The radial variable as a solution of the orbit equations, such as the Binet equation, depends on an angular variable and is characterized by a coupling parameter. In application to orbits, the identity transformation $r = \rho$ means $r(\theta; \lambda_a) = \rho(\tilde{\theta}; \lambda_b)$, which occurs when $\theta \to \tilde{\theta}$. The angular transformation $\tilde{\theta} = \theta + \theta_0$ where $-2\pi < \theta_0 < 2\pi$ causes a rotation of a given orbit $\rho(\tilde{\theta}; \lambda_b) = r(\theta; \lambda_a) = r(\tilde{\theta} - \theta_0; \lambda_a)$ about the center of force by θ_0. For instance, the cardioid orbit $r = r_0 \cos^2(\theta/2)$ in a potential with power $a = -3$ maps into $\rho = r_0 \sin^2(\tilde{\theta}/2)$ by a rotation $\tilde{\theta} = \theta + \pi$. This example belongs to the self-dual pair $(-3, -3)$ via $\eta = 1$. In this regard, we argue that the identity transformation includes rotations about the center of forces. Of course, the rotation with $\theta_0 = 0$ is the bona fide identity transformation.*

Remark 7. *Suppose two circular orbits pass through the center of attraction. It is known that the attraction is an inverse fifth-power force. If the radii of the two circles are the same, then the inverse fifth-power force is self-dual under a rotation. If the radii of the two circles are different, the two orbiting objects must possess different masses. A map between two circles with different radius, passing through the center of the same attraction, is precluded from possible links for the self-dual pair $(-4, -4)$ by the mass invariance requirement (iv).*

Remark 8. *If $C_1 < 0$ in (25), either r or ρ must be negative contrary to our initial assumption. However, when we consider the mapping of orbits, as we do in Remark 6, we recognize that there is a situation where the angular change $\theta \to \tilde{\theta}$ induces $\rho(\tilde{\theta}; \lambda_b) = -r(\theta; \lambda_a) = r(\theta; -\lambda_a)$. For instance, consider an orbit given by a conic section $r = p/(1 + e\cos\theta)$ where $p > 0$ and $-1/e < \cos\theta \le 1$. If $e > 1$, then it is possible to find $\tilde{\theta}$ such that $-1 \le \cos\tilde{\theta} < -1/e$ by $\theta \to \tilde{\theta}$. Consequently the image of the given orbit is $\rho(\tilde{\theta}; p) = r(\tilde{\theta}; p) = -r(\theta; p) < 0$. Certainly the result is unacceptable. The latus rectum p is inversely proportional to λ_a. Hence in association with the sign change in coupling $\lambda_a \to \lambda_b = -\lambda_a$, we are able to obtain a passable orbit $\rho(\tilde{\theta}, -p) = r(\tilde{\theta}; -p) = -r(\theta; -p) > 0$. The orbit mapping of this type cannot be achieved by a rotation. To include the situation like this in the space transformation, we formally introduce the inversion,*

$$\mathfrak{R}_i: \qquad r \to -\rho, \tag{30}$$

and treat it as if the case of $C_1 = -1$. Then we interpret the negative sign of the radial variable as a result of a certain change in the angular variable θ involved in the orbital equation by associating it with a sign change in coupling so that both r and ρ remain positive. If $\mu = 1$, the inversion causes no change in time, mass, energy, and angular momentum, but entails, as is apparent from (27), a change in coupling,

$$\lambda_a \to \lambda_b = (-1)^a \lambda_a. \tag{31}$$

The inversion set Δ_1 with $C_1 = -1$ and $\mu = 1$, denoted by Δ_i, is partially qualified as a duality transformation. The reason why Δ_i is "partially" qualified is that it is admissible only when a is an integer. Notice that $(-1)^a$ appearing in (31) is a complex number unless a is an integer. As λ_a and λ_b are both assumed to be real numbers, a must be integral. Having said so, in the context of the inversion, we need a further restriction on a. The sign change in coupling is induced by the inversion only when a is an odd number. Since Δ_i is not generally reducible to the identity set Δ_0, it is non-trivial.

2.3.2. Class II

Class II is the set of proper (traditional) dual pairs. Equation (20) of the second combination can be expressed as

$$\mathfrak{C}_2: \qquad \eta = 2/(a+2) \quad \text{with} \quad b = -2a/(a+2), \qquad (a \ne -2). \tag{32}$$

which implies that a pair $(a, b) = (a, -2a/(a+2))$ is linked by η_a when $a \neq -2$. The above operation $\mathfrak{C}_2$ may as well be given by

$$\mathfrak{C}_2' : \quad \eta = (b+2)/2 \quad \text{with} \quad a = -2b/(b+2), \quad (b \neq -2), \tag{33}$$

which means a pair $(a, b) = (-2b/(b+2), b)$ linked via $\eta = (b+2)/2$. Another expression for $\mathfrak{C}_2$ is

$$\mathfrak{C}_2'' : \quad \eta = (b+2)/2, \quad \text{with} \quad (a+2)(b+2) = 4, \quad (a \neq -2, b \neq -2), \tag{34}$$

which is a version of what Needham [22,23] calls the Kasner–Arnol'd theorem for dual forces. If $a \neq 0$ and $b \neq 0$,

$$\eta = 2/(a+2) = (b+2)/2 = -b/a, \quad (a \neq -2, b \neq -2), \tag{35}$$

from which follows that to every (a, b) via η_a there corresponds (b, a) via η_a^{-1} if $a \neq 0, -2$. If $|a| \ll 1$, then $b \approx -a$ and $(a, b) \approx (a, -a)$. Hence $(0,0) = \lim_{a \to 0}(a, -a)$ via η_+, which overlaps with $(0,0) = \lim_{a \to 0}(a, a)$ of Class I in the limit but differs in approach. In the above η_a stand for η with a fixed a.

In this case, the transformation functions of (12) can be written as $f(\rho) = C_a \rho^{\eta_a}$ and $g(\rho) = \mu C_a^2 \eta_a^2 \rho^{2\eta_a - 2}$ where $C_a = C_{\eta_a}$. Here we choose $\mu = 1$ by the reason stated in Remark 4. The change of radial variable (6) and the change of time derivative (7) become, respectively,

$$\mathfrak{R}_a : \quad r = C_a \rho^{\eta_a}, \tag{36}$$

and

$$\mathfrak{T}_a : \quad (dt/d\varphi) = C_a^2 \eta_a^2 \rho^{2\eta_a - 2}(ds/d\varphi). \tag{37}$$

Equation (21) of the second combination, associated with $\{\mathfrak{R}_a, \mathfrak{T}_a\}$, yields the coupling-energy exchange operation,

$$\mathfrak{E}_a : \quad \lambda_b = -C_a^2 \eta_a^2 E_a, \quad E_b = -C_a^{a+2} \eta_a^2 \lambda_a, \quad (a \gtrless -2). \tag{38}$$

The time scaling has been chosen so as to preserve the mass invariance (11),

$$\mathfrak{M}_0 : \quad m_b = m_a = m, \tag{39}$$

and the scale change in the angular momentum follows from (13) with η_a,

$$\mathfrak{L}_a : \quad L_b = \eta_a L_a. \tag{40}$$

Now we see that each of the sets $\Delta_a = \{\mathfrak{C}_a, \mathfrak{R}_a, \mathfrak{T}_a, \mathfrak{E}_a, \mathfrak{M}_0, \mathfrak{L}_a\}$ preserves the form-invariance of the action (4) with a power potential. The form-invariance warrants that $X(a, b)\Delta_a = \Delta_b$. Hence system B is dual to system A with respect to Δ_a. Let $\Delta_\pm = \{\Delta_a; a \gtrless -2\}$. The set Δ_+ links $a > -2$ and $b > -2$ of (a, b), whereas Δ_- relates $a < -2$ to $b < -2$. No Δ_a links $a \gtrless -2$ to $b \lessgtr -2$. Hence there is no pair (a, b) consisting of $a \gtrless -2$ and $b \lessgtr -2$.

Remark 9. *Class II consists of proper dual pairs (a, b) linked by $\Delta_\pm$, which have been widely discussed in the literature [10,18,22–24,36,40]. Here a and b are distinct except for two self-dual pairs, $(0,0)$ via η_+ and $(-4, -4)$ via η_-.*

Remark 10. *Note that the time transformation (37) is not integrable unless the time-dependence of the space variable (i.e., the related orbit) is specified.*

Remark 11. *The scale factor C_1 appeared in Case I was dimensionless. A space transformation of (12) for a given value of η_a contains a constant C_{η_a} which has a dimension of $r^{a/(a+2)}$. Let $C_{\eta_a} = C_a d_a$ where C_a and d_a are a dimensionless magnitude and the dimensional unit of C_{η_a},* *respectively. Use of an appropriate scale transformation which is admissible as seen in Case I enables C_a to reduce to unity. More over, the dimensional unit may be suppressed to $d_a = 1$. Therefore, if desirable, the space transformation (36) may simply be written as $r = \rho^{1/a}$ without altering physical contents.*

Remark 12. *Let (a,b) be a dual pair satisfying the relation $(a+2)(b+2) = 4$. Then the left element $(a, \ )$ of (a,a) maps via (a,b) into $(b, \ )$, and the right element $(\ ,a)$ into $(\ ,b)$. Hence the self-dual pair (a,a) can be taken by (a,b) to the self-dual pair (b,b). Schematically,*

$$(a,a) \xrightarrow{(a,b)} (b,a) \xrightarrow{(a,b)} (b,b).$$

We call $((a,a),(b,b))$ a grand dual pair.

2.4. Graphic Presentation of Dual Pairs

A dual pair (a,b) is presented as a point in a two-dimensional $a - b$ plane as shown in Figure 1. All self-dual pairs (a,a) of Class I are on a dashed straight line $a = b$ denoted by η_1. Every dual pair (a,b) of Class II is shown as a point on two branches $\eta_\pm$ of a hyperbola described by the equation $(a+2)(b+2) = 4$ of (34). The graph for Class II is similar to the one given by Arnol'd for dual forces [10].

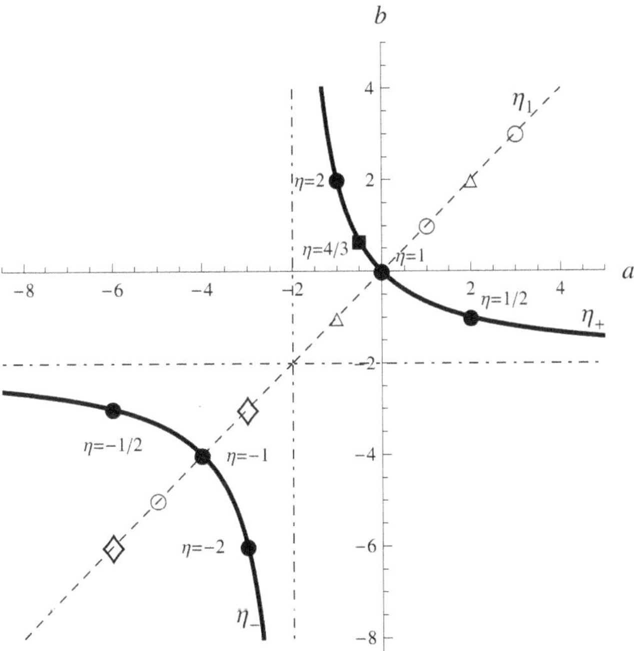

Figure 1. The solid line shows the allowed combinations of dual pairs (a,b) of power laws. The dashed line indicates the symmetry axis $(a,b) \leftrightarrow (b,a)$. The bullets show the only dual pairs where both a and b are integers representing the Newton–Hook duality. The square represents the duality pair discussed in Section 4.4.

Among the dual pairs of Class I, there are pairs (a,a) linked by scale transformations (inclusive of rotations), which cover all real a, and those (a,a) related by the inversion, which are defined only when a is an odd number. In this regard, every pair (a,a), occupying a single point on η_1, plays multiple roles. While the pairs linked by scale transformations admissible for all real values of a form a continuous line η_1 indicated by a dashed line, those pairs linked by the inversion appear as discrete points on η_1 and are indicated by circles.

The hyperbola representing all pairs of Class II has its center at $(-2,-2)$, transverse axis along $b = a$, and asymptotes on the lines $a = -2$ and $b = -2$. The bullets indicate all pairs (a,b) via $\eta_\pm$ with integral a's; namely, $(-1,2)$ via $\eta = 2$, $(0,0)$ via $\eta = 1$, $(-3,-6)$ via $\eta = -2$, and $(-4,-4)$ via $\eta = -1$. There are no integer pairs other than those listed above in Class II. The square represents the dual pair $(-1/2, 2/3)$ to be discussed in Section III D. On the branch of η_+, a dual pair (a,b) via η_+ and its inverse pair (b,a) via η_+^{-1} are symmetrically located about the transverse axis η_1. Since both (a,b) and (b,a) signify that system A and system B are dual to each other, the curves $\eta_\pm$ have redundancy in describing the $A - B$ duality. An example is the Newton–Hooke duality for which two equivalent pairs $(-1,2)$ via $\eta = 2$ and $(2,-1)$ via $\eta = 1/2$ appear in symmetrical positions on η_+.

We notice that there are two special points on the graph. They are the intersections of η_1 and $\eta_\pm$; namely, $(0,0)$ with $\eta = 1$, and $(-4,-4)$ with $\eta = \pm 1$. The former is an overlapping point of η_1 and η_+ where $\eta = 1$. The latter is like an overhead crossing of η_1 and η_- where the pair belonging to η_1 is linked by a transformation with $\eta = 1$ while the one belonging to η_- is linked with $\eta = -1$.

In approaching the crossing of η_1 and η_+, the pair $(0,0)$ at $\eta_1 = 1$ has a limiting behavior as $(0,0) = \lim_{\varepsilon \to 0}(\pm\varepsilon, \pm\varepsilon)$, while $(0,0)$ at $\eta_+ = 1$ behaves like $(0,0) = \lim_{\varepsilon \to 0}(\pm\varepsilon, \mp\varepsilon)$ via $\eta = 1$. As has been mentioned earlier, $(a,b) = (a-1, b-1)_F$. However, the counterpart of $(0,0)$ is not exactly equal to $(-1,-1)_F$. The potential corresponding to the inverse force $F \sim 1/r$ is $V \sim \ln r$. Thus, it is more appropriate to put symbolically $(-1,-1)_F = (\ln, \ln)$. Yet, $(0,0) \neq (\ln, \ln)$. Consider $V_a(r) = \lambda_a r^\varepsilon$. For ε small, $V_a(r) \approx \lambda_a(1 + \varepsilon \ln r)$, which gives rise to the force $F \approx \kappa/r$ where $\kappa = \lambda_a \varepsilon$. As long as κ can be treated as finite, $(\varepsilon, -\varepsilon) \approx (-1,-1)_F$. Chandrasekhar [18] excluded $(-1,-1)_F$ from the list of dual pairs on physical grounds. We exclude $(\ln, \ln)$ because the logarithmic potential, being not a power potential, lies outside our interest.

By analyzing Corollaries and Propositions in the Principia, Chandrasekhar [18] pointed out that Newton had found not only the Newton–Hooke dual pair but also the self-dual pairs $(2,2)$, $(-1,-1)$ and $(-4,-4)$. He also mentioned that $(-3,-6)$ was not included in the Prinpicia. For a integral, there are only two grand dual pairs $((-1,-1),(2,2))$ and $((-3,-3),(-6,-6))$. In Figure 1, $(2,2)$ and $(-1,-1)$ are marked with triangles on η_1, while $(-3,-3)$ and $(-6,-6)$ are marked with diamonds on η_1.

2.5. Classical Orbits

Here we discuss the orbital behaviors for the dual pairs in relation with energy and coupling.

First, we consider self-dual pairs (a,a) of Class I. If an effective shifted potential is defined by $U^{eff}(r) = U(r) + L^2/(2mr^2)$, the space transformation $r = C_1\rho$ induces

$$U_a^{eff}(r) = \lambda_a r^a + \frac{L_a^2}{2mr^2} - E_a, \quad \Rightarrow \quad U_b^{eff}(\rho) = C_1^{a+2}\lambda_a \rho^a + \frac{L_a^2}{2m\rho^2} - C_1^2 E_a, \tag{41}$$

resulting in self-dual pairs (a,a) for any real a. The space transformation includes scale transformations $r = C_1\rho$ with $C_1 > 0$, identity transformation $r = \rho$ (inclusive of rotations), and inversion formally defined by $r = -\rho$.

Statement 1. *System A and system B linked by a scale transformation are physically identical but described in different scale. Typically an orbit of system A maps to an orbit of system B similar in shape but different in scale.*

Statement 2. *In the limit $C_1 \to 1$, the two orbits become congruent (identical) to each other. Any self-dual pair (a,a) due to a scale transformation is reducible to a trivial pair (a,a) linked by the identity transformation. However, in dealing with the orbital behaviors, we have to look into the angular dependence of radial variables by allowing the identity transformation $r = \rho$ to contain*

$r(\theta) = \rho(\tilde{\theta}) = r(\tilde{\theta} - \theta_0)$ with $\theta \to \tilde{\theta} = \theta + \theta_0$, which represents a rotation of a given orbit about the center of force by θ_0.

The inversion $r \to -\rho$ entails $\lambda_b = (-1)^a \lambda_a$, as is apparent from (41). If a is an even number, the sign change in coupling does not occur. Hence the inversion for even a cannot properly be defined and must be precluded. Only when a is odd, the inversion is meaningful. However, we have to notice that orbits in a potential with $a > 0$ are all bounded if $\lambda_a > 0$ and all unbounded if $\lambda_a < 0$. Under the inversion, the sign of λ_a changes, so that a bound orbit with $E_a > 0$ is supposed to go to an unbounded orbit with $E_b = E_a > 0$. It is uncertain whether there are such examples to which the inversion works.

Statement 3. *If a is a negative odd number, under the inversion, an orbit in an attractive (repulsive) potential maps to an orbit in a repulsive (attractive) potential, keeping the energy unchanged.*

In the Principia, Newton proved that if an orbit passing through the center of attraction is a circle then the force is inversely proportional to the fifth-power of the distance from the center (Corollary I to Proposition VII). From Corollary I of Proposition VII and other corollaries in the Principia Chandrasekhar [18] shows in essence that if an object moves on a circular orbit under centripetal attraction emanating from two different points on the circumference of the circle then the forces from the two points exerted on the orbiting object are of the same inverse fifth-power law. Then he suggests, in this account, that the inverse fifth power law of attraction is self-dual for motion in a circle. In contrast to Chandrasekhar's view on the self-dual pair $(-4, -4)$, we maintain that $(-4, -4)$ can be understood as a member of Class I and Class II. The circular orbit in an attractive potential $V_a(r) = \lambda_a r^{-4}$, which occurs when $E_a = 0$, can be described by the equation $r = 2R_a \cos\theta$ where $R_a = \sqrt{-\lambda_a m / (2L_a^2)}$ is the radius of the circle and $-\pi/2 < \theta < \pi/2$ is the range of θ. The scale transformation $r = C_1 \rho$ with $C_1 > 0$ converts the orbit equation into $\rho = 2R_b \cos\theta$ where $R_b = R_a/C_1$. Apparently it is consistent with the requirements $L_b = L_a$ and $\lambda_b = C_1^{-2}\lambda_a$ of (41). Thus, the radius of the circle is rescaled while the center of force is fixed at the origin and the range of θ is unaltered. The inverse fifth-power law of attraction may be viewed as self-dual under a scale change for motion in a circle. If the identity map $r = \rho$ may include a rotation $r(\theta) \to \rho(\tilde{\theta}) = r(\tilde{\theta} - \theta_0)$, then $\rho(\tilde{\theta}) = 2R\cos(\tilde{\theta} - \theta_0)$ with the angular range $-\pi/2 + \theta_0 < \tilde{\theta} < \pi/2 + \theta_0$. In particular, if $\theta_0 = \pi$, then $\rho(\tilde{\theta}) = -2R\cos(\tilde{\theta})$ with $\pi/2 < \tilde{\theta} < 3\pi/2$. The circular orbit maps into itself, though rotated about the center of force. In this sense, the inverse fifth-power law of attraction is self-dual under a rotation for motion in a circle. In much the same way, the inverse fifth-power force, whether attractive or repulsive, may be considered as self-dual under a scale change and a rotation for motion in any other orbits. Hence the self-dual pair $(-4, -4)$ linked by the scale transformation (including rotations) is a member of Class I. The same self-dual pair $(-4, -4)$ has another feature as a member of Class II which will be discussed in Remark 13.

Secondly, we consider dual pairs (a, b) of Class II.

All dual pairs (a, b) of Class II are subject to the proper dual transformation Δ_{II}. The members a and b of each pair obey the Kasner–Arnol'd formula $(a + 2)(b + 2) = 4$, and are related via $\eta = 2/(a + 2)$ (or $\eta = -b/a$ if $a \neq 0$). These dual pairs belong to branch η_+ if $a > -2$, and branch η_- if $a < -2$.

Now the space and time transformations $r = C_a \rho^{2/(a+2)}$ and $(dt/d\varphi) = C_\eta^2 \eta^2 \rho^{-2a/(a+2)}(ds/d\varphi)$ induce the energy-coupling exchange,

$$\lambda_b = -C_\eta^2 \eta^2 E_a, \qquad E_b = -C_\eta^{a+2}\eta^2 \lambda_a, \tag{42}$$

where $C_\eta > 0$ and $a \neq -2$. Hence the effective shifted potential transforms as

$$U_a^{eff}(r) = \lambda_a r^a + \frac{L_a^2}{2mr^2} - E_a, \quad \Rightarrow \quad U_b^{eff}(\rho) = -C_\pm^2 \eta_\pm^2 E_a \rho^b + \frac{\eta_\pm^2 L_a^2}{2m\rho^2} + C_\pm^{a+2}\eta_\pm^2 \lambda_a \tag{43}$$

where $a \neq -2$ and $b = -2a/(a+2)$.

The two equations in (42) are not simply to exchange the roles of energy and coupling. They also provide a useful relation between E_a and E_b. In general E_a depends on λ_a. So we let $E_a = \mathcal{E}_a(\lambda_a)$, and invert it as $\lambda_a = \mathcal{E}_a^{-1}(-\lambda_b/\eta^2 C_\eta^2)$ with the help of the first equation of (42). Substitution of this into the second equation of (42) yields

$$E_b = -\eta^2 C_\eta^{a+2} \mathcal{E}_a^{-1}\left(-\frac{\lambda_b}{\eta^2 C_\eta^2}\right). \tag{44}$$

which shows that E_b depends on E_a through the coupling λ_a.

Statement 4. *For a dual pair (a, b) of Class II, if the coupling dependence of E_a is explicitly known, then E_b can be determined by (44), and vice versa.*

From (42) there follow four possible mapping patters,

$$
\begin{aligned}
(0) \qquad & (E_a = 0, \lambda_a \gtrless 0) \implies (E_b \lessgtr 0, \lambda_b = 0) \\
(1) \qquad & (E_a > 0, \lambda_a < 0) \implies (E_b > 0, \lambda_b < 0) \\
(2) \qquad & (E_a < 0, \lambda_a < 0) \implies (E_b > 0, \lambda_b > 0) \\
(3) \qquad & (E_a > 0, \lambda_a > 0) \implies (E_b < 0, \lambda_b < 0) \\
(4) \qquad & (E_a < 0, \lambda_a > 0) \implies (E_b < 0, \lambda_b > 0)
\end{aligned}
$$

In the above, pattern (0) implies that any zero energy orbit of system A goes to a rectilinear orbit of system B with no potential. Patterns (1)–(4) imply that any positive energy orbit of system A, regardless of the sign of λ_a, maps to an orbit of system B with a coupling $\lambda_b < 0$, and any negative energy orbit of system A, independent of λ_a, maps to an orbit of system B with a coupling $\lambda_b > 0$.

The dual pairs (a, b) of Class II can be grouped into those on η_+ and those on η_-. Furthermore, the pairs of the first group can be divided into two parts for $\eta_+ > 1$ and $0 < \eta_+ < 1$. If we let $\eta_+^>$ denote the part for $\eta_+ > 1$, then $\eta_+^> = \{-b/a | -2 < a < 0, b > 0\}$. Similarly, let $\eta_+^<$ denote the part for $0 < \eta_+ < 1$. Then $\eta_+^< = \{-b/a | a > 0, -2 < b < 0\} = \{-a/b | -2 < a < 0, b > 0\}$. Thus, $\eta_+^< = [\eta_+^>]^{-1}$. It is sufficient to consider the set $\eta_+^>$. The same can be said for the second group on η_-. We take up only the set $\eta_-^> = \{-b/a | -4 < a < -2, b < -4\}$.

For the case of $\eta_+^>$, $\lambda_a > 0\,(<0)$ implies a repulsion (attraction), while $\lambda_b > 0\,(<0)$ means an attraction (repulsion). There are no negative energy orbits in a repulsive potential with $\lambda_a > 0$ and in an attractive potential with $\lambda_b > 0$. For $\eta_-^>$, both $\lambda_a > 0$ and $\lambda_b > 0$ are repulsive, and both $\lambda_a < 0$ and $\lambda_b < 0$ are attractive. In any repulsive potential with $\lambda_a > 0$ or $\lambda_b > 0$, no negative energy orbits are present. Pattern (4) is not physically meaningful. Taking these features of potentials into account, we can restate the implication of the relations in (42) as follows.

Statement 5. *Under the proper duality transformation Δ_{II}, if $-2 < a < 0$ (i.e., $b > 0$), then any positive energy orbit in the potential of system A, whether attractive or repulsive, maps to an orbit in a repulsive potential of system B, and any negative energy (bound) orbit maps to a positive energy (bound) orbit in an attractive potential. If $a > 0$ (i.e., $-2 < b < 0$), then the above situations are reversed. If $-4 < a < -2$ (i.e., $b < -4$), then any positive orbit in an attractive potential maps to a positive orbit under attraction, any negative bound orbit in an attractive potential maps to a positive orbit under repulsion, and any positive orbit under repulsion maps to a negative bound orbit in an attractive potential. Even for the case where $a < -4$ (i.e., $-4 < b < -2$), the mapping patterns are the same as those for $-4 < a < -2$. In all cases, zero energy orbits map to force-free rectilinear orbits.*

This is a modified version of Needham's statement made in supplementing the Kasner–Arnol'd theorem [22,23].

Remark 13. *The pair* $(-4, -4)$ *has another feature as a point on* η_-, *that is, as a member of Class II. From (42), it is obvious that* $\lambda_b = 0$ *for the circular zero energy orbit. Hence the duality transformation* Δ_{II} *maps the orbit into a force-free rectilinear orbit. According to Statement 5, any positive energy orbit must map to an orbit in an attractive potential, and any negative energy orbit maps to an orbit in a repulsive potential. Therefore, the self-dual pair* $(-4, -4)$ *Newton established is not a member of Class II. It must be* $(-4, -4)$ *on* η_1, *belonging to Class I.*

In what follows, we make remarks on the Newton–Hooke pairs and related self-dual pairs.

Remark 14. *Statement 5 applies to the pair* $(-1, 2)$. *The mapping patterns (0)–(3) works in going from the Newton system with* $a = -1$ *to the Hooke system with* $b = 2$. *Namely, (0) the zero energy orbit of the attractive Newton system maps to a rectilinear orbit; (1) a positive unbound orbit of the attractive Newton system maps to a positive unbound orbit of the repulsive Hooke system; (2) a negative energy bound orbit of the attractive Newton system maps to a positive energy bound orbit of the attractive Hooke system; and (3) a positive unbound orbit of the repulsive Newton system maps to a negative unbound orbit of the repulsive Hooke system. Since there are no negative orbits for the repulsive Newton system and the attractive Hooke system, pattern (4) is irrelevant.*

Remark 15. *In view of the orbit structure, we study in more detail the mapping process from the Newton system to the Hooke system. As is well-known, for the motion in the inverse-square force, the orbit equation in polar coordinates has the form,*

$$r = \frac{p}{1 + e\cos\theta}, \tag{45}$$

where p *is the semi-latus rectum, e the eccentricity. The orbit is of conic sections and the origin of the coordinates is at the focus closest to the pericenter of the orbit. The angle* θ *is between the position of the orbiting object and the direction to the pericenter located at* $r = r_{min}$ *and* $\theta = 0$. *The semi-latus rectum, the semi-major axis, and the eccentricity of the orbit are determined by* $p = -L_a^2/(m\lambda_a)$, $\bar{a} = -\lambda_a/(m|E_a|)$, *and* $e = \sqrt{1 + (2L_a^2 E_a/m\lambda_a^2)}$, , *respectively. If the inverse square force is attractive, i.e., if* $\lambda_a < 0$, *then* $\bar{a} > 0$, $p > 0$, *and* $1 > \cos\theta > -1/e$. *If repulsive, i.e., if* $\lambda_a > 0$, *then* $\bar{a} < 0$, $p < 0$ *and* $-1 < \cos\theta < -1/e$.

(i) For the bound motion, $E_a < 0$, $e < 1$ *and* $p = \bar{a}(1 - e^2) > 0$. *The Equation (45) describes an elliptic orbit with semi-major axis* $\bar{a}$ *and eccentricity e. Apparently,* $r_{min} = \bar{a}(1 - e)$. *For the duality mapping, a more suited choice is the orbit equation expressed in terms of the eccentric anomaly* ψ,

$$r = \bar{a}(1 - e\cos\psi), \tag{46}$$

which may be put in the form,

$$r = \bar{a}\Big\{(1 + e)\cos^2(\psi/2) + (1 - e)\sin^2(\psi/2)\Big\}. \tag{47}$$

Here ψ *is related to the polar angle* θ *by* $\tan(\theta/2) = [(1 + e)/(1 - e)]^{1/2}\tan(\psi/2)$. *Since* $r = C_2\rho^2$, *use of (47) leads to*

$$\rho = \Big[\alpha^2\cos^2(\psi/2) + \beta^2\sin^2(\psi/2)\Big]^{1/2}, \tag{48}$$

where

$$\alpha = \sqrt{\bar{a}(1 + e)/C_2}, \qquad \beta = \sqrt{\bar{a}(1 - e)/C_2}. \tag{49}$$

Let $\rho = \sqrt{u^2 + v^2}$ *in cartesian coordinates, and let*

$$u = \alpha\cos(\psi/2), \qquad v = \beta\sin(\psi/2). \tag{50}$$

Then it is clear that the trajectory drawn by ρ is given as an ellipse,

$$\frac{u^2}{\alpha^2} + \frac{v^2}{\beta^2} = 1, \tag{51}$$

with semi-major axis α and semi-minor axis β, centered at the origin of the $u - v$ plane. It is obvious that $\rho_{min} = \sqrt{\bar{a}(1-e)/C_2}$ is the semi-minor axis of the ellipse on the $u - v$ plane. The above calculation shows that the elliptic Kepler orbit with semi-major axis $\bar{a}$ and eccentricity e maps to an ellipse with semi-major axis $\alpha = \sqrt{\bar{a}(1+e)/C_2}$ and eccentricity $\epsilon = \sqrt{2e/(1+e)}$. The semi-major and semi-minor axes of the resultant ellipse depend on the scaling factor C_2. With different values of C_2, a Kepler ellipse of eccentricity e is mapped to ellipses of different sizes having a common eccentricity ϵ. In general, the resultant ellipse having eccentricity ϵ is not similar to the Kepler orbit with eccentricity e. If $e = 0$, then $\epsilon = 0$. Namely, a circular orbit of radius $\bar{a}$ under an inverse-square force maps to a circle with radius $\alpha = \sqrt{\bar{a}/C_2}$. With a particular scale $C_2 = 1/\sqrt{\bar{a}}$, the mapped circle is congruent to the original orbit. In the limit $e \to 1$, the Kepler orbit becomes a parabola with $E_a = 0$, which maps to a force-free rectilinear orbit described by $(u, v) = (\rho, 0)$.

(ii)   If $E_a > 0$, then $e > 1$ and $\bar{a} > 0$ for $\lambda_a < 0$. The semi-latus rectum in (45) must be modified as $p = \bar{a}(e^2 - 1) > 0$. Again $\cos\theta < -1/e$. The orbit is a branch of a hyperbola with semi-major axis $\bar{a}$ and eccentricity e. The center of attraction is at the interior focus of the branch, so that $r_{min} = \bar{a}(e - 1)$. In much the same fashion that the eccentric anomaly is used in (46), we introduce a parameter ψ related to the angle θ by $\tan(\theta/2) = [(e+1)/(e-1)]^{1/2}\tanh(\psi/2)$. Here $\cosh\psi > 1/e$. Now the orbit equation in parametric representation is

$$r = \bar{a}(e\cosh\psi - 1), \tag{52}$$

which may further be written as

$$r = \bar{a}\left\{(e-1)\cosh^2(\psi/2) + (e+1)\sinh^2(\psi/2)\right\}, \tag{53}$$

whose minimum occurs when $\psi = 0$. Correspondingly, $\rho = \sqrt{r/C_2}$ is expressed as

$$\rho = \left[\alpha^2\cosh^2(\psi/2) + \beta^2\sinh^2(\psi/2)\right]^{1/2}, \tag{54}$$

where

$$\alpha = \sqrt{\bar{a}(e-1)/C_2}, \qquad \beta = \sqrt{\bar{a}(e+1)/C_2}. \tag{55}$$

Hence $\rho_{min} = \sqrt{\bar{a}(e-1)/C_2}$. Letting

$$u = \alpha\cosh(\psi/2), \qquad v = \beta\sinh(\psi/2), \tag{56}$$

we obtain $\rho = \sqrt{u^2 + v^2}$ and the equation for a hyperbola having two branches,

$$\frac{u^2}{\alpha^2} - \frac{v^2}{\beta^2} = 1, \tag{57}$$

which has the semi-major axis $\alpha = \sqrt{\bar{a}(e-1)/C_2}$ and the eccentricity $\epsilon = \sqrt{2e/(e-1)}$. Thus, the positive energy orbit in the attractive inverse potential, given by a branch of the hyperbola, maps to a positive energy orbit given by either branch of a hyperbola whose center coincides with the center of the repulsive Hooke force.

(iii)   For a repulsive potential with $\lambda_a > 0$ such as the repulsive Coulomb potential, the orbit Equation (45) describing a hyperbola holds true insofar as $E_a > 0$, i.e., $e > 1$. Since $p = -L_a^2/(m\lambda_a) < 0$ for $\lambda_a > 0$, the semi-lotus rectum must be replaced by $\tilde{p} = -p$. At the same time, the angular variable has to be changed from θ to $\tilde{\theta}$ where $\cos\theta < -1/e$ and $\cos\tilde{\theta} > -1/e$.

The conversion of the hyperbolic Equation (45) for the attractive potential to the hyperbolic equation for the repulsive potential,

$$\tilde{r} = \frac{\tilde{p}}{1 + e\cos\tilde{\theta}}, \tag{58}$$

is indeed the inversion process mentioned in Remark 8. Since (45) and (58) have the same form, we can follow the procedure given in (ii) to show that under $\tilde{r} = \sqrt{\rho/C_2}$ the positive energy orbit in the repulsive inverse potential, given by a branch of the hyperbola, maps to a negative energy orbit given by either branch of a hyperbola whose center coincides with the center of the repulsive Hooke force.

Remark 16. *In connection with Remark 14, we look at the self-dual pairs $(-1, -1)$ and $(2, 2)$ which do not belong to Class II. Apparently the two pairs are closely related to each other via the Newton–Hooke pair $(-1, 2)$, so as to form a grand dual pair $((-1, -1), (2, 2))$. As they are both on η_1, each of them is self-dual under scale changes and rotations. In addition, $(-1, -1)$ is self dual under the inversion. From (iii) of Remark 15, it is clear that due to the inversion the orbit equation takes the form (45). There the angular range for $\tilde{\theta}$ is $\theta_e < \tilde{\theta} < 2\pi - \theta_e$ where $\theta_e = \cos^{-1}(-1/e)$. Hence the resultant orbit has the center of orbit at the exterior focus. This means that a hyperbolic orbit in attraction with the center of force at the interior focus maps to the conjugate hyperbola in repulsion with the center of force at the exterior focus. In contrast, any rotation maps a hyperbolic orbit under attraction (repulsion) into a hyperbolic orbit under attraction (repulsion). In summary, the inversion maps a hyperbolic orbit under attraction into a hyperbolic orbit under repulsion, whereas any rotation takes a hyperbolic orbit under attraction (repulsion) to a hyperbolic orbit under attraction (repulsion). According to Chandrasekhar's book [18], what Newton established for $(-1, -1)$ and $(2, 2)$ are that the attractive inverse square force law is dual to the repulsive inverse square force law, and that the repulsive linear force law is dual to itself. Thus, we are led to a view that Newton's $(-1, -1)$ is due to the inversion and their $(2, 2)$ is due to a rotation. Finally we wish to point out that by the mapping patterns (1) and (3) of $(-1, 2)$ a hyperbolic orbit of the attractive Newton system, whether attractive or repulsive, maps to a hyperbolic orbit of the repulsive Hooke system. In other words, the pair of forces $(attraction, repulsion)$ for $(-1, -1)$ goes to the pair of force $(repulsion, repulsion)$ for $(2, 2)$ with the help of $(-1, 2)$. This is compatible with the assertion that Newton's two self-dual pairs form the grand dual pair $((-1, -1), (2, 2))$ via $(-1, 2)$.*

2.6. Classical Energy Formulas

We have used the energy-coupling exchange relations,

$$\mathfrak{E}: \quad E_b = -\eta^2 C^{a+2}\lambda_a, \quad \lambda_b = -\eta^2 C^2 E_a, \tag{59}$$

as essential parts of the power-duality operations. They demand primarily that the roles of energy and coupling be exchanged. Using these relations, we can also derive energy formulas which enable us to determine the energy value of one system from that of the other when two systems are power-dual to each other.

In general E_a depends on λ_a, L_a and possibly other parameters. So let the energy function be $E_a = \mathcal{E}(\lambda_a, L_a, w_a)$ where w_a represents those additional parameters. Then we pull λ_a out from the inside of $\mathcal{E}$ as

$$\lambda_a = \mathcal{E}^{-1}(E_a, L_a, w_a). \tag{60}$$

Now we insert this coupling parameter λ_a into the first equation of (59). Substituting the second relation $E_a = -\lambda_b/(\eta^2 C^2)$ and the angular momentum transformation $L_a = L_b/\eta$ to the right-hand side of (60), we can convert the first relation of (59) into an energy formula,

$$E_b(\lambda_b, L_b, w_b) = -\eta^2 C^{a+2}\mathcal{E}^{-1}(-\lambda_b/(\eta^2 C^2), L_b/\eta, w_a(w_b)). \tag{61}$$

Thus, if E_a is known, then E_b can be determined without solving the equations of motion for system B. By making an appropriate choice of C, the value of λ_b may be specified by the second relation of (59).

Alternatively, let us combine the two relations in (59) by eliminating the constant C to get another energy formula,

$$E_b = -\eta^2 \lambda_a \left(-\frac{\lambda_b}{\eta^2 E_a} \right)^{1/\eta}. \tag{62}$$

This formula can be rearranged to the symmetric form,

$$\left[4(a+2)^{-2} |\lambda_a|^{-2/(a+2)} |E_a| \right]^a = \left[4(b+2)^{-2} |\lambda_b|^{-2/(b+2)} |E_b| \right]^b. \tag{63}$$

Note that the signs of the energies and coupling constants are related via (59). See also the four patterns discussed in *Statement 4* above.

When the parameters w contained in E_a are invariant, that is, $w_a = w_b$, under the duality operations, the last equation suggests that there is some positive function $\mathcal{F}(L, w)$, independent of λ_a and λ_b, such that

$$|E_a(\lambda_a, L_a, w)| = \frac{(a+2)^2}{4} |\lambda_a|^{2/(a+2)} \left\{ \mathcal{F}\left(\sqrt{2/(a+2)}\, L_a, w \right) \right\}^{1/a}, \tag{64}$$

$$|E_b(\lambda_b, L_b, w)| = \frac{(b+2)^2}{4} |\lambda_b|^{2/(b+2)} \left\{ \mathcal{F}\left(\sqrt{2/(b+2)}\, L_b, w \right) \right\}^{1/b}, \tag{65}$$

where $L = \sqrt{(a+2)/2}\, L_a = \sqrt{(b+2)/2}\, L_b$. If such a function is specified for E_a by (64), then E_b can be determined by (65) with the sign to be obtained via (59). Notice that (65) is useful as an energy formula to find E_b only when E_a has the form of (64).

Remark 17. *As an example, let us consider the Newton–Hooke dual pair for which* $(a, b) = (-1, 2)$, $\eta = -b/a = 2$ *and* $r = C\rho^2$. *Let system A be consisting of a particle of mass m moving around a large point mass $M \gg m$ under the influence of the gravitational force with* $\lambda_a = -GmM < 0$. *Let system B be an isotropic harmonic oscillator with* $\lambda_b = \frac{1}{2}m\omega^2 > 0$. *Then, as the exchange relations of (59) demand, $E_a < 0$ and $E_b > 0$. Hence the orbits of the two systems are bounded. This means that the Newton–Hooke duality occurs only when both systems are in bound states.*

Suppose the total energy of the particle is given in the form,

$$E_a = \frac{L_a^2}{2mr_{min}^2} + \frac{\lambda_a}{r_{min}} = \mathcal{E}(\lambda_a, L_a, r_{min}), \tag{66}$$

where r_{min} is the minimum value of the radial variable r and $\lambda_a = -GmM$. Then we obtain the inverse function,

$$\lambda_a = \mathcal{E}^{-1}\left(-\lambda_b/4C^2, L_a, r_{min} \right) = -\frac{L_a^2}{2mr_{min}} - \frac{\lambda_b r_{min}}{4C^2}. \tag{67}$$

With this result, the Formula (60) immediately leads to the energy of the Hooke system in the form,

$$E_b = \frac{L_b^2}{2m\rho_{min}^2} + \lambda_b \rho_{min}^2 \tag{68}$$

where $L_b = 2L_a$ and $\rho_{min} = \sqrt{r_{min}/C}$. Although λ_b may be interpreted as Hooke's constant, its detailed form $\frac{1}{2}m\omega^2$ cannot be determined by the energy formula. Noticing that E_a is a constant, we let $\kappa = \sqrt{-2mE_a}$. If we choose $C = m\omega/(2\kappa)$, then we have $\lambda_b = \frac{1}{2}m\omega^2$ from the second relation of (59). With the same choice of C, we have $m\omega\rho_{min}^2 = 2\kappa r_{min}$.

Suppose the energy of system A is alternatively given in the form,

$$E_a = -\frac{2\pi^2 m\lambda_a^2}{(J + 2\pi L_a)^2}, \tag{69}$$

where J is the radial action variable, $J = \oint dr\, p_r$, or more explicitly,

$$J = 2\int_{r_{min}}^{r_{max}} dr \sqrt{2m\left(E - \frac{\lambda_a}{r} - \frac{L^2}{2mr^2}\right)}, \tag{70}$$

which is a constant of motion. Let E_a of (69) be put into the form given via (64) then we may identify

$$\mathcal{F}\left(\sqrt{2/(a+2)}\, L_a, J\right) = \left[J/(2\pi) + (\sqrt{2}(L_a)/\sqrt{2}\right]^2/(2m). \tag{71}$$

From this follows

$$\left\{\mathcal{F}\left(\sqrt{2/(b+2)}\, L_b, J\right)\right\}^{1/2} = [J/(2\pi) + (L_b)/2]/\sqrt{2m} \tag{72}$$

Since the first relation of (59) indicates that $E_b > 0$ for $\lambda_a < 0$, the relation (65) together with $\lambda_b = \frac{1}{2}m\omega^2$ results in

$$E_b = (\omega/2\pi)(2J + 2\pi L_b), \tag{73}$$

which is an energy expression of the Hooke system obtainable from the Hamilton-Jacobi equation.

2.7. Generalization to Multi-Term Power Laws

In the following, on a parallel with Johnson's treatment [36], we examine how the duality can be realized with a sum of power potentials (i.e., a multi-term potential) in the present framework.

Let the potential V_a be a sum of N distinct power potentials as

$$V_a(r) = \sum_{i=1}^{N} \lambda_{a_i} r^{a_i}, \quad a_i > -2, \quad (a_i \neq a_j \quad \text{for} \quad i \neq j) \tag{74}$$

where λ_{a_i} is the coupling constant of the i-th sub-potential in V_a. Then $\mathfrak{R}$ and $\mathfrak{T}$ take the shifted potential in (16) to

$$gU_a(r) = \sum_{i=1}^{N} \lambda_{a_i} C^{a_i+2} \eta^2 \rho^{2\eta-2+a_i\eta} - C^2 \eta^2 \rho^{2\eta-2} E_a. \tag{75}$$

Let us pick one of the terms in the sum in (75), say, the $i = k$ term, and make its exponent zero by letting

$$\eta = \eta_k = \frac{2}{a_k + 2}, \quad a_k > -2, \tag{76}$$

where η is k-dependent. If the exponent of the $i = k'$ term, instead of the $k \neq k'$ term, is made vanishing, then η is to be given in terms of $a_{k'}$ where $a_{k'} \neq a_k$. Since $k = 1, 2, \ldots, N$, there are N possible choices of η. Thus, it is appropriate to write η in (76) with the subscript k as η_k. Apparently, η_k is a possible one of $\{\eta_1, \eta_2, \ldots, \eta_N\}$. Let the operations $\mathfrak{R}$ and $\mathfrak{T}$ for $\eta = \eta_k$ be denoted by $\mathfrak{R}_k$ and $\mathfrak{T}_k$, respectively.

For the remaining potential terms $(i \neq k)$ and the energy term in (75), we rename the exponents of ρ as

$$b_k = -\frac{2a_k}{a_k + 2}, \quad b_i = \frac{2(a_i - a_k)}{a_k + 2}, \quad i \neq k, \tag{77}$$

which can easily be inverted to express a_k and a_i in terms of b_k and b_i in the same form. These relations are equivalent to the conditions on the exponents,

$$\mathfrak{C}_k: \quad (a_k + 2)(b_k + 2) = 4, \quad (a_i - a_k)(b_i - b_k) = a_i b_i. \tag{78}$$

From (77) there also follows $b_i > -2$ for all i if $a_i > -2$ for all i. The first relation of (78) leads to alternative but equivalent expressions of η in (76),

$$\eta_k = -\frac{b_k}{a_k} = \frac{b_k + 2}{2} = \frac{2}{a_k + 2}. \tag{79}$$

To $\mathfrak{R}_k$ and $\mathfrak{T}_k$, we have to add two more operations,

$$\mathfrak{L}_k: \quad L_{b_k} = \eta_k L_{a_k}, \tag{80}$$

and

$$\mathfrak{W}_k: \quad \lambda_{b_k} = -C^2 \eta_k^2 E_a, \quad E_{b_k} = -\eta_k^2 C^{a_k+2} \lambda_{a_k}, \quad \text{and} \quad \lambda_{b_i} = \eta_k^2 C^{a_i+2} \lambda_{a_i}, \quad i \neq k. \tag{81}$$

Then, we express the shifted potential of (75) in the new notation as

$$g U_a(r) = V_{b_k}(\rho) - E_{b_k} = U_{b_k}(\rho) \tag{82}$$

where

$$V_{b_k}(\rho) = \sum_{i=1}^{N} \lambda_{b_i} \rho^{b_i}. \tag{83}$$

The set of operations $\Delta_k = \{\mathfrak{R}_k, \mathfrak{T}_k, \mathfrak{C}_k, \mathfrak{L}_k, \mathfrak{W}_k\}$ transforms the radial action of the A system into

$$W_\rho(E_{b_k}) = \int_{I_\varphi} d\varphi \, (ds/d\varphi) \left\{ \frac{m}{2} (ds/d\varphi)^{-2} \left(\frac{d\rho}{d\varphi} \right)^2 - \frac{L_{b_k}^2}{2m\rho^2} - U_{b_k}(\rho) \right\}. \tag{84}$$

Thus, we find the duality between the A-system and B_k-system with respect to Δ_k. Again, this duality is only one of the N dualities; there are N pairs of dual systems, (a_k, b_k) for $k = 1, 2, ..., N$.

3. Power-Duality in the Semiclassical Action

The power-duality argument made for the classical action in Section 2 can easily be carried over to the semiclassical action. In semiclassical theory the power-duality is a relationship between two quantum systems which are not mutually interacting. In studying such a relationship, there are two distinct approaches; one is to pay attention to a reciprocal relation between two systems, and the other to pursue a deeper connection between the quantum states of two systems (see Remark 18). Our power-duality argument is of the former approach, taking reciprocity as a heuristic guiding. Special care will have to be exercised though, when dealing with the quantum structure of each system.

3.1. Symmetry of the Semiclassical Action

The action in semiclassical theory is of the form, $W = \int dq\, p$, which is Hamilton's characteristic function and essentially the same as that in (2). The semiclassical action for the radial motion reads

$$W = \int dr \sqrt{2m\left(E - V(r) - \hbar^2 L^2/(2mr^2)\right)}. \tag{85}$$

Here the classical angular momentum L is replaced by $\hbar L$. Customarily the semiclassical angular momentum (divided by $\hbar$) of (85) is given by the Langer-modified form,

$$L = \ell + (D-2)/2, \quad \ell = 0, 1, 2, \dots \tag{86}$$

if it is defined in D dimensions. Let us write the semiclassical action for system A as

$$W_a = \int dr \sqrt{-2m\left[\hbar^2 L_a^2/(2mr^2) + U_a(r)\right]} \tag{87}$$

where $U_a(r) = V_a(r) - E_a$. After the change of variable $r = f(\rho)$, the action (87) of system A becomes

$$W_a = \int d\rho \sqrt{-2m\left[\hbar^2 L_a^2 g/(2mf^2) + gU_a(f)\right]}, \tag{88}$$

where $f' = dr/d\rho$ and $g = f'^2$. The following substitutions

$$\mathfrak{R}: \quad f(\rho) = C\rho^\eta, \tag{89}$$

$$\mathfrak{L}: \quad L_a = L_b/\eta, \tag{90}$$

$$gU_a = U_b, \tag{91}$$

lead the action (88) to

$$W_b = \int d\rho \sqrt{-2m\left[\hbar^2 L_b^2/(2m\rho^2) + U_b(\rho)\right]}, \tag{92}$$

which is taken as the action for system B. Here we have assumed $m_a = m_b = m$ (i.e., $\mu = 1$). We shall also assume that two mutually power-dual systems are by definition in the same dimensions (i.e., $D_a = D_b = D$).

Only when the potential of system A is a power potential, $U_b(\rho)$ in (92) can be brought to the form $V_b(\rho) - E_b$. The change of variable $\mathfrak{R}: r = C\rho^\eta$ with the choice $\mathfrak{C}_2: \eta = 2/(a+2)$ gives $g(\rho) = \eta^2 C^2 \rho^{-a\eta}$. Hence, for $V_a(f) = \lambda_a C^a \rho^{a\eta}$, we have $g(\rho)V_a(f) = \eta^2 C^{a+2}\lambda_a$ and $gE_a = \eta^2 C^2 E_a \rho^b$ where $b = -a\eta = -2a/(a+2)$. After performing the energy-coupling exchange,

$$\mathfrak{E}: \quad \lambda_a = -E_b/(\eta^2 C^{a+2}), \quad E_a = -\lambda_b/(\eta^2 C^2), \tag{93}$$

we obtain

$$g(\lambda_a r^a - E_a) = \lambda_b \rho^b - E_b. \tag{94}$$

In effect, under the operation of g, the following transformations have taken place,

$$gV_a(r) \to -E_b, \quad gE_a \to -V_b(\rho), \tag{95}$$

where $V_a = \lambda_a r^a$ and $V_b(\rho) = \lambda_b \rho^b$.

In this manner, transforming the action W_a of (87) to W_b of (92) by the duality operations, we have $W_a = W_b$, that is,

$$\int dr \sqrt{2m(E_a - \lambda_a r^a) - \hbar^2 L_a^2/r^2} = \int d\rho \sqrt{2m(E_b - \lambda_b \rho^b) - \hbar^2 L_b^2/\rho^2}. \tag{96}$$

It is also apparent that $W_a = X(a,b)W_b$ with $\zeta_a = r$ and $\zeta_b = \rho$. Thus, we see that the semiclassical action (85) is form-invariant under the set of duality operations, $\{\mathfrak{R}, \mathfrak{L}, \mathfrak{C}, \mathfrak{E}\}$.

Although we have presented in the above the power-duality features of the semi-classical action similar to those in the classical case, we have not taken account of the possibility that the angular momentum L is a discretely quantized entity given in terms of the angular quantum number $\ell = 0, 1, 2, \ldots$ by (86). It is natural to expect that the operation $\mathfrak{L}: L_b = \eta L_a$ of (90) implies the equality,

$$\ell_b + (D_b - 2)/2 = \eta \ell_a + \eta (D_a - 2)/2. \tag{97}$$

In addition, if we demand that $\ell_a = 0$ corresponds to $\ell_b = 0$, then (97) can be separated into two equalities,

$$\ell_b = \eta \ell_a, \qquad D_b = \eta(D_a - 2) + 2. \tag{98}$$

Either (97) or (98) suggests that the allowed values of ℓ_b differs from those of ℓ_a unless $\eta = 1$. This means that the condition $\ell = 0, 1, 2, \ldots$ in (86) cannot be imposed on system A and system B at the same time. Although the transformations in (97) and (98) are invertible, they cannot preserve the Langer-form (86) of the angular momentum in the two systems. In other words, they are not reciprocal relations between the two systems. Insofar as operation $\mathfrak{L}$ implies the equalities (97), the semiclassical action with the Langer modification is not form-invariant under the set of operations $\{\mathfrak{R}, \mathfrak{L}, \mathfrak{C}, \mathfrak{E}\}$. Then, we may have to draw a conclusion that the power-duality valid in the classical action breaks down in the semiclassical action due to the quantized angular momentum term.

In the above we have observed that the power-duality is incompatible with the angular quantization. By the same token, the energy-coupling relations of $\mathfrak{E}$ in (93) may have to be examined. In the semiclassical action, the energy E and the coupling λ may be treated as parameters. However, the implication of the exchange relations in (93) becomes ambiguous after quantization. It is not clear whether E_a in (93) is one of the energy eigenvalues of system A or it represents the energy spectrum of the system. As an aid of clarification, we study one of the energy formulas resulting from combining the two relations in (93),

$$E_b = -\eta^2 \lambda_a \left(-\frac{\lambda_b}{\eta^2 E_a} \right)^{1/\eta}, \tag{99}$$

which has been given in Section 2 as a classical energy formula. To see if it will work in quantum mechanics, let us employ, e.g., the Coulomb–Hooke duality, the quantum counterpart of the Newton–Hooke duality, and test (99). We assume that E_a and E_b in (99) represent the spectra of system A and system B, respectively. According to (99), the energy spectrum E_b of the hydrogen atom with the Coulomb coupling $\lambda_b = -e^2$ is expected to follow from the spectrum E_a of the three-dimensional isotropic harmonic oscillator with frequency $\omega = \sqrt{2\lambda_a/m}$. For this pair of systems, $(a,b) = (2,-1)$ and $\eta = -b/a = 1/2$. Given $E_a(n_r, \ell_a) = \hbar\omega(2n_r + \ell_a + 3/2)$ with $n_r = 0, 1, 2, \ldots$ and $\ell_a = 0, 1, 2, \ldots$, the Formula (99) immediately yields $E_b = -(me^4/2\hbar^2)(n_r + \ell_a/2 + 3/4)^{-2}$. Here $n = n_r + \ell_a/2 + 3/4 = 3/4, 5/4, 7/4, \ldots$. The result is not the energy spectrum of the hydrogen atom that is commonly known. Evidently, a naive application of the energy Formula (99) fails at the level of angular quantum numbers. By contrast, if we consider the states of a four-dimensional oscillator which possess $\ell_a = 0, 2, 4, \ldots$, then $n = n_r + \ell_b + 1 = 1, 2, 3, \ldots$ via $\ell_b = \ell_a/2$, which matches the principal quantum number of the hydrogen atom. In other words, the energy Formula (99) suggests that the spectrum of the hydrogen atom can be composed of "half the states" of the four dimensional isotropic harmonic oscillator (to be more precise, the set $\{\ell_a = 0, 2, 4, \ldots, D_a = 4\}$ for the oscillator and the set $\{\ell_b = 0, 1, 2, \ldots, D_b = 3\}$ for the H-atom are in one-to-one correspondence). The relation between the oscillator in four dimensions and the hydrogen atom in three dimensions is not reciprocal in (99). The alternative scheme is not the Coulomb–Hooke duality that we pursue (see Remark 18). The Coulomb–Hooke duality in quantum mechanics will be discussed again in Section 4.3.

In an effort to make the power duality meaningful in semiclassical theory, we shall take a view that the power duality is basically a classical notion. Accordingly, for the duality discussions, all physical objects such as L, E and λ, should be treated as classical entities, i.e., continuous parameters. Then we consider quantization as a process separate from the duality operations. The duality is a classical feature of the relation between two systems, whereas quantization is associated with the micro-structures of each system. None of duality operations can dictate how the quantum structure of each system should be. The equality of (93) which is compatible with reciprocity must not imply the non-reciprocal equality of (97). It is necessary to dissociate duality operations from quantization. Technically, we deal only with those continuous parameters for the duality discussions, and replace them as a post duality-argument activity by appropriately quantized counterparts when needed for characterizing each quantum system. From this view, the power duality of the semiclassical action has already been established at the equality (96) with follow-up substitutions $L_a = \ell_a + (D-2)/2$, $(\ell_a = 0, 1, 2, ..)$ and $L_b = \ell_b + (D-2)/2$, $(\ell_b = 0, 1, 2, ...)$. It is helpful to introduce the dot-equality $\doteq$ to signify the equality amended by substitutions of quantized entities. The power-duality of the semiclassical action in the amended version may be exhibited by

$$\int \mathrm{d}r \sqrt{2m(E_a - \lambda_a r^a) - \hbar^2(\ell_a + (D-2)/2)^2/r^2}$$
$$\doteq \int \mathrm{d}\rho \sqrt{2m(E_b - \lambda_b \rho^b) - \hbar^2(\ell_b + (D-2)/2)^2/\rho^2}. \tag{100}$$

3.2. The Semiclassical Energy Formulas

In the preceding section, we have adopted the Coulomb–Hooke duality to test (99), and failed. However, it should be recognized that if the energy spectrum of the three dimensional radial oscillator is given in the form $E_a(n_r, L_a) = \hbar\omega(2n_r + L_a + 1)$ without requiring $L_a = \ell_a + 1/2$, then the energy formula (99) together with $L_a = 2L_b$ yields $E_b(n_r, L_b) = -(me^4/2\hbar^2)(n_r + L_b + 1/2)^{-2}$ which reduces to the desired Coulomb spectrum $E_b(\nu, L_b) = -(me^4/2\hbar^2)(n_r + \ell_b + 1)^{-2}$ after ad hoc substitution of $L_b = \ell_b + 1/2$ with $\ell_b \in \mathbb{N}_0$. So long as L, E and λ are treated as continuous parameters, the energy formula (99) derived from the exchange relations (93) should work for semiclassical systems provided that those parameters are eventually replaced by their quantum counterparts.

In semiclassical theory, the bound state energy E_a of system A can be evaluated by carrying out the integration on the left-hand side of (96) between two turning points. Namely, we calculate for E_a the integral

$$J_a = 2 \int_{r'}^{r''} \mathrm{d}r \sqrt{2m(E_a - \lambda_a r^a) - \hbar^2 L_a^2/r^2}, \tag{101}$$

where r' and r'' are the turning points of the orbit where the integrand vanishes. The quantity J_a is indeed an action variable defined for a periodic motion by $\oint \mathrm{d}q\, p$. It is a constant depending on E_a, λ_a, and L_a. By letting it be a constant N_a multiplied by $2\pi\hbar$,

$$J_a(E_a, \lambda_a, L_a) = 2\pi\hbar N_a, \tag{102}$$

and solving (102) for E_a, we obtain the classical bound state energy as a function of parameters λ_a, L_a and N_a,

$$E_a = E_a(\lambda_a, L_a, N_a). \tag{103}$$

Once the classical energy E_a of system A is given in terms of λ_a, L_a and N_a, when system A and system B are power-dual to each other, we can determine the energy E_b of system B, with the help of the operations $\mathfrak{L}$ and $\mathfrak{E}$, as a function of λ_b, L_b and N_b. Since $W_a = W_b$ as shown in (96), it is obvious that $N_a = N_b$. As the former equality is a consequence of the duality operations, so is the latter equality. Hence the equality $N_a = N_b$

is a consequence but not a part of duality operations. So, we let $N = N_a = N_b$. With the energy function (103), the semiclassical energy formula stemming from (99) is

$$E_b(\lambda_b, L_b, N) = -\eta^2 \lambda_a \left(-\frac{\lambda_b}{\eta^2 E_a(\lambda_a, L_b/\eta, N)} \right)^{1/\eta}, \tag{104}$$

which can be rearranged as the classical case in the following form

$$|E_a(\lambda_a, L_a, N)| = \frac{1}{4}(a+2)^2 |\lambda_a|^{2/(a+2)} \left\{ \mathcal{F}\left(\sqrt{2/(a+2)}\, L_a, N \right) \right\}^{1/a} \tag{105}$$

$$|E_b(\lambda_b, L_b, N)| = \frac{1}{4}(b+2)^2 |\lambda_b|^{2/(b+2)} \left\{ \mathcal{F}\left(\sqrt{2/(b+2)}\, L_b, N \right) \right\}^{1/b} \tag{106}$$

where $\mathcal{F}(L, N)$ is a function common to both systems. The signs for both energy relations are determined as in the classical case via the signs of the coupling constants, i.e., $\mathrm{sgn}\, E_a = -\mathrm{sgn}\, \lambda_b$ and $\mathrm{sgn}\, E_b = -\mathrm{sgn}\, \lambda_a$.

Alternatively, expressing an explicit form of the energy function (103) by $\mathcal{E}(\lambda_a, L_a, N)$ as

$$E_a = \mathcal{E}(\lambda_a, L_a, N_a), \tag{107}$$

and inverting (107) to take λ_a out, we have

$$\lambda_a = \mathcal{E}^{-1}(E_a, L_a, N). \tag{108}$$

Then we use the angular momentum transformation $\mathfrak{L}$ of (90) and the energy-coupling exchange relations $\mathfrak{E}$ of (93) to write down the bound state energy formula for E_b as

$$E_b(\lambda_b, L_b, N) = -\eta^2 C^{a+2} \mathcal{E}^{-1}(-\lambda_b/(\eta^2 C^2), L_b/\eta, N), \tag{109}$$

which is essentially the same as the energy formula (104).

To convert the classical energy E_a in (103) to a quantum spectrum, we replace the parameters L_a and N by their corresponding quantized entities. The angular momentum is quantized in the Langer form $L_a \doteq \ell_a + (D-2)/2$. The Wentzel–Kramers–Brillouin (WKB) quantization formula for the radial motion,

$$\oint \mathrm{d}r\, p_r = 2\pi\hbar\left(n_r + \frac{1}{2} \right), \quad n_r = 0, 1, 2, \ldots \tag{110}$$

asserts that

$$N \doteq n_r + 1/2, \quad n_r = 0, 1, 2, \ldots \tag{111}$$

Substitution of the Langer-modified angular momentum (86) and the WKB quantization (111) in the classical energy function of (103) yields the energy spectrum,

$$E_a(n_r, \ell_a) \doteq E_a(\lambda_a, \ell_a - 1 + D/2, n_r + 1/2), \tag{112}$$

where $n_r = 0, 1, 2, \ldots$ and $\ell_a = 0, 1, 2, \ldots$ Similarly, after substitution of the Langer form (86) to L_b and the WKB quantization (111) to N, the semiclassical energy formula (109) leads to the energy spectrum of system B,

$$E_b(n_r, \ell_a) \doteq -\eta^2 C^{a+2} \mathcal{E}^{-1}\left(-\lambda_b/(\eta^2 C^2), (\ell_b - 1 + D/2)/\eta, n_r + 1/2 \right) \tag{113}$$

where $n_r = 0, 1, 2, \ldots$ and $\ell_b = 0, 1, 2, \ldots$.

3.3. A System with a Non-Integer Power Potential and Zero-Angular Momentum

As a simple but non-trivial example, we study a non-integer power potential system with $L^2 = 0$ (see Remark 22). Let system A be the case. Bound states of system A occur only when (i) $\lambda_a < 0$, $a < 0$ with $E_a < 0$ or (ii) $\lambda_a > 0$, $a > 0$ with $E_a > 0$. The integral (101) with $L_a = 0$, denoted $J_a(E_a, \lambda_a, 0)$, is reducible to a beta function under either condition (i) or (ii). Suppose system A be under condition (i). Then it goes to a beta function as

$$J_a(E_a, \lambda_a, 0) = M(E_a, \lambda_a) \int_0^1 dz\, z^{-\frac{a+2}{2a}-1}(1-z)^{\frac{3}{2}-1} = M(E_a, \lambda_a) B\left(-\frac{a+2}{2a}, \frac{3}{2}\right) \quad (114)$$

where we have let $z = (E_a/\lambda_a)r^{-a}$ and $M(E_a, \lambda_a) = \sqrt{-2m\lambda_a/a^2}(E_a/\lambda_a)^{(a+2)/2a}$. As in (102), we express the right-hand side of (114) by the parameter N as

$$M(E_a, \lambda_a) B\left(-\frac{a+2}{2a}, \frac{3}{2}\right) = 2\pi\hbar N, \quad (115)$$

which we solve for E_a to find the energy function $E_a = \mathcal{E}(\lambda_a, 0, N)$,

$$E_a = -(-\lambda_a)^{\frac{2}{a+2}}\left(\frac{\sqrt{2m}}{\hbar|a|\pi} B\left(-\frac{a+2}{2a}, \frac{3}{2}\right)\right)^{-\frac{2a}{a+2}} N^{\frac{2a}{a+2}}. \quad (116)$$

Now the WKB condition (111) yields the energy spectrum of system A,

$$E_a(n) \doteq -(-\lambda_a)^{\frac{2}{a+2}}\left(\frac{\sqrt{2m}}{\hbar|a|\pi} B\left(-\frac{a+2}{2a}, \frac{3}{2}\right)\right)^{-\frac{2a}{a+2}}\left(n+\frac{1}{2}\right)^{\frac{2a}{a+2}}, \quad (117)$$

where $n = 0, 1, 2, \ldots$. The bound state energy spectrum of system B can be independently calculated in a similar fashion, and the WKB quantization (111), separately applied to system B, will lead to a spectrum similar to but different from the spectrum of system A in (117). Insofar as system B is power-dual to system A, the bound state energy spectrum of system B can be obtained via the formula (113). Inverting the λ_a dependent function (116), we obtain

$$\lambda_a = \mathcal{E}^{-1}(E_a, 0, N) = -(-E_a)^{(a+2)/2}\left(\frac{\sqrt{2m}}{\hbar|a|\pi} B\left(-\frac{a+2}{2a}, \frac{3}{2}\right)\right)^a N^{-a}. \quad (118)$$

Utilizing this inverted function and the WKB condition (111) in the energy Formula (113), we arrive at the energy spectrum of system B,

$$E_b(n) \doteq \lambda_b^{\frac{2}{b+2}}\left(\frac{\sqrt{2m}}{\hbar|b|\pi} B\left(\frac{1}{b}, \frac{3}{2}\right)\right)^{-\frac{2b}{b+2}}\left(n+\frac{1}{2}\right)^{\frac{2b}{b+2}}, \quad n = 0, 1, 2, \ldots \quad (119)$$

which is independent of the arbitrary constant C appearing in (109). In the above, we have also changed a to b by using the relations, $a = -2b/(b+2)$ and $\eta a = -b$. Apparently, the spectrum (119) is very similar in form with the spectrum of system A in (109) but is not identical. The relations (93) suggest that $E_b > 0$ for $\lambda_a < 0$ and $\lambda_b > 0$ for $E_a < 0$. Hence system B has bound states with $E_b > 0$ only when $b > 0$. This means that system B is under condition (ii) and that the energy spectrum (119) is for the case where $\lambda_b > 0$, $b > 0$ with $E_b > 0$. In particular, if $V_a(r) = \lambda_a/\sqrt{r}$ with $\lambda_a < 0$, the spectrum resulting from (117) is

$$E_{a=-1/2}(n) \doteq \frac{\lambda_a}{2}\left(-\frac{m\lambda_a}{\hbar^2}\right)^{1/3}\left(n+\frac{1}{2}\right)^{-2/3}, \quad n = 0, 1, 2, \ldots. \quad (120)$$

For the dual partner potential $V_b(\rho) = \lambda_b \rho^{2/3}$ with $\lambda_b > 0$, the spectrum follows from (119) as

$$E_{b=2/3}(n) \doteq 2\lambda_b \left(\frac{8\hbar^2}{9m\lambda_b} \right)^{1/4} \left(n + \frac{1}{2} \right)^{1/2}, \quad n = 0, 1, 2, \dots . \tag{121}$$

3.4. Duality in SUSY Semiclassical Formulas

Let us begin this section with a brief comment on the semiclassical quantization in supersymmetric quantum mechanics (SUSYQM). In SUSYQM, there are semiclassical quantization formulas similar to WKB's. A unified form of them for a radial motion is

$$\int_{r'}^{r''} dr \sqrt{2m(E - \Phi^2(r))} = \pi\hbar \left(\nu + \frac{1}{2} + \frac{\Delta}{2} \right), \quad \nu = 0, 1, 2, \dots , \tag{122}$$

defined for the partner Hamiltonians $H_\pm$. In (122), E is the eigenvalue of $H_\pm$, and $\Phi(r)$ is the superpotential which is a solution of the Riccati equation in the form

$$\Phi^2(r) \pm \frac{\hbar}{\sqrt{2m}} \frac{d\Phi(r)}{dr} - V(r) - \frac{\hbar^2(L^2 - \frac{1}{4})}{2mr^2} = 0 \tag{123}$$

where $V(r)$ is a potential function, r' and r'' denote the turning points defined by $\Phi^2(r') = E = \Phi^2(r')$ with $r' \le r''$, and $L = \ell + (D-2)/2$ with $\ell = 0, 1, 2, \dots$. There, Δ is the Witten index whose values are $\Delta = \pm 1$ for good SUSY and $\Delta = 0$ for broken SUSY (SUSY stands for supersymmetry. If $H_\pm$ are the partner Hamiltonians, then $\mathrm{spec}(H_-) \setminus \{0\} = \mathrm{spec}(H_+)$ for good SUSY, and $\mathrm{spec}(H-) = \mathrm{spec}(H_+)$ for broken SUSY). The quantization condition for good SUSY was found by Comtet, Bandrauk, and Campell [59]. The broken SUSY case and the general formulation of the form (123) were derived by Eckhardt [60] and independently by Inomata and Junker [61]. It is known that both the Comtet–Bandrauk–Campbell (CBC) formula for good SUSY and the Eckhardt–Inomata–Junker (EIJ) condition for broken SUSY yield the exact energy spectra for many shape-invariant potentials. For detail, see reference [62].

Now we wish to study the power-duality in SUSY semiclassical action on the left-hand side of (122) only for the H_- case. Let us write the action of system A as

$$W_a = \int_{r'}^{r''} dr \sqrt{2m(E_a - \Phi_a^2(r))}, \tag{124}$$

where E_a is the eigenvalue of H_-. Suppose the superpotential in (124) has the form,

$$\Phi_a(r) = \epsilon \sqrt{\lambda_a} r^{a/2} - \frac{\hbar}{\sqrt{2m}} \frac{\mu_a}{r}, \tag{125}$$

where $\epsilon = \pm 1$ and a in the shoulder of r is an arbitrary real number. The potential term appearing in the SUSY semiclassical action (124) is the squared-superpotential rather than the usual potential $V(r)$. For the superpotential (125), it is

$$\Phi_a^2(r) = \lambda_a r^a + \lambda_{a'} r^{a'} + \frac{\hbar^2 \mu_a^2}{2mr^2}, \tag{126}$$

where

$$a' = (a-2)/2, \quad \lambda_{a'} = -\epsilon \hbar \mu_a \sqrt{2\lambda_a/m}. \tag{127}$$

Then we have

$$\Phi_a^2(r) - \frac{\hbar}{\sqrt{2m}} \frac{d\Phi_a(r)}{dr} = \lambda_a r^a + \left(1 + \frac{a}{4\mu_a} \right) \lambda_{a'} r^{a'} + \frac{\hbar^2 \mu_a(\mu_a - 1)}{2mr^2}. \tag{128}$$

Evidently, $\Phi_a(r)$ of (125) satisfies the Riccati Equation (123) with a two-term power potential,

$$V_a(r) = \lambda_a r^a + (1 + a/(4\mu_a))\lambda_{a'} r^{a'}, \tag{129}$$

provided that

$$a' = (a-2)/2, \qquad \mu_a = L_a + 1/2. \tag{130}$$

In (129), a is arbitrary but a' is dependent on a as given by the first condition of (127). If both a and a' are assumed to be independent and arbitrary, the superpotential of the form (125) cannot be a solution of the Riccati equation. The quantity on the left-hand side of (128) is a SUSY effective potential, denoted by $V_a^{(-)}(r)$, that belongs to the Hamiltonian H_-. It is related to $V_a(r)$ of (129) by

$$V_a^{(-)}(r) = \Phi_a^2(r) - \frac{\hbar}{\sqrt{2m}}\frac{d\Phi_a(r)}{dr} = V_a(r) + \frac{\hbar^2(L^2 - \frac{1}{4})}{2mr^2}. \tag{131}$$

The superpotential (125) works for the radial oscillator and the hydrogen atom in a unified manner as it contains the two as special cases:

(1) Radial harmonic oscillator with $a = 2$, $a' = 0$, $\lambda_a = \frac{1}{2}m\omega^2$, $\lambda_{a'} = -\hbar\omega\mu_a$, $\mu_a = L_a + \frac{1}{2}, \epsilon = 1$:

$$\Phi_a(r) = \sqrt{\frac{m}{2}}\omega r - \frac{\hbar}{\sqrt{2m}}\frac{\mu_a}{r}, \tag{132}$$

$$V_a^{(-)}(r) = \frac{1}{2}\omega^2 r^2 + \frac{\hbar^2\mu_a(\mu_a - 1)}{2mr^2} - \hbar\omega(\mu_a + 1/2), \tag{133}$$

$$E_a - \Phi_a^2 = (E_a + \hbar\omega\mu_a) - \frac{1}{2}m\omega^2 r^2 - \frac{\hbar^2\mu_a^2}{2mr^2}. \tag{134}$$

The CBC quantization of (122) with $\Delta = -1$ yields

$$E_a = 2\hbar\omega\nu, \qquad \nu \in \mathbb{N}_0, \tag{135}$$

which corresponds to the energy spectrum in quantum mechanics,

$$E_a^{QM}(\nu, \ell) = E_a + \hbar\omega\mu_a = \hbar\omega(2\nu + \ell + D/2 - 1/2), \tag{136}$$

if $\mu_a = L_a + 1/2 = \ell + D/2 - 1/2$ with $\ell \in \mathbb{N}_0$.

(2) Hydrogen atom with $a = 0$, $a' = -1$, $\epsilon = 1$, $\lambda_a = me^4/(2\hbar^2\mu_a^2)$, $\lambda_{a'} = -e^2$, $\mu_a = L_a + \frac{1}{2}$:

$$\Phi_a(r) = \frac{\sqrt{2m}}{2\hbar\mu_a}e^2 - \frac{\hbar}{\sqrt{2m}}\frac{\mu_a}{r}, \tag{137}$$

$$V_a^{(-)}(r) = -\frac{e^2}{r} + \frac{\hbar^2\mu_a(\mu_a - 1)}{2mr^2} + \frac{me^4}{2\hbar^2\mu_a^2}, \tag{138}$$

$$E_a - \Phi_a^2 = \left(E_a - \frac{me^4}{2\hbar^2\mu_a^2}\right) + \frac{e^2}{r} - \frac{\hbar^2\mu_a^2}{2mr^2}. \tag{139}$$

The CBC result is

$$E_a = E_a^{QM}(\nu, \ell) + me^4/(2\hbar^2\mu_a^2) = -\frac{me^4}{2\hbar^2(\nu + \ell + D/2 - 1/2)^2} + \frac{me^4}{2\hbar^2(\ell + D/2 - 1/2)^2}, \tag{140}$$

where $\nu, \ell \in \mathbb{N}_0$.

Next we change the radial variable r by

$$\mathfrak{R}: \qquad r = f(\rho) = C\rho^{\eta}, \qquad \Leftrightarrow \qquad \rho = f^{-1}(r) = C^{-1/\eta}r^{1/\eta}. \tag{141}$$

and let the system described by the new variable be system B. Upon application of (141), the action W_a of (124) transforms to

$$W_b = \int_{\rho'}^{\rho''} d\rho \sqrt{2mf'^2(E_a - \Phi_a^2)}, \tag{142}$$

where $f' = df(\rho)/d\rho$ and

$$E_a - \Phi_a^2 = E_a - \lambda_a r^a - \lambda_{a'} r^{a'} - \frac{\hbar^2 \mu_a^2}{2mr^2}, \qquad a' = (a-2)/2. \tag{143}$$

Since $f'^2 = \eta^2 C^2 \eta^{2\eta-2}$,

$$f'^2(E_a - \Phi_a^2) = \eta^2 C^2 E_a \rho^{2\eta-2} - \eta^2 C^{2+a}\lambda_a \rho^{a\eta+2\eta-2} - \eta^2 C^{2+a'}\lambda_{a'}\rho^{a'\eta+2\eta-2} - \frac{\hbar^2 \eta^2 \mu_a^2}{2m\rho^2}. \tag{144}$$

If there is such a parameter η that $f'^2(E_a - \Phi_a^2)$ takes the form,

$$E_b - \Phi_b^2 = E_b - \lambda_b \rho^b - \lambda_{b'}\rho^{b'} - \frac{\hbar^2 \mu_b^2}{2m\rho^2}, \tag{145}$$

with

$$b' = (b-2)/2, \tag{146}$$

then the action is form-invariant under (141) and reciprocal, that is, $W_a = W_b$ and $W_a = \hat{X}(a,b)W_b$. In the $\hat{X}(a,b)$-operation, we have temporarily let $r = \xi_a$ and $\rho = \xi_b$. We have also assumed that $\hat{X}(a,b)$ takes $b' = (b-2)/2$ to $a' = (a-2)/2$. Furthermore, (145) together with (146) implies that the new superpotential $\Phi_b(\rho)$ has the same form as that of $\Phi_a(r)$ in (125), namely,

$$\Phi_b(r) = \epsilon\sqrt{\lambda_b}r^{b/2} - \frac{\hbar}{\sqrt{2m}}\frac{\mu_b}{r}. \tag{147}$$

If this were the case, we could establish the general power-duality of the action (124) with the superpotential (125). Unfortunately there is no way to transform system A with an arbitrary power a to system B satisfying the conditions (145) and (146). Therefore, with the superpotential (125), we are unable to demonstrate in a general term the power-dual symmetry in SUSY semiclassical quantization. To our knowledge, no qualified superpotential supporting the general power-duality in SUSY semiclassical action has ever been reported.

Although we have to give up pursuing the general power-duality, we may find cases where duality occurs within the present scheme. For a dual symmetry, the form-invariance of the superpotential $\Phi(r)$ is not an essential requirement, but it is necessary that $f'^2(E_a - \Phi_a^2(r))$ is reducible to the form $E_b - \Phi_b^2$ under the transformation $r = f(\rho) = C\rho^{\eta}$. There are two options for η to reduce the left-hand side of (144) to the form of (145) under different conditions than (146). Namely,

$$(i) \qquad \eta = 2/(a+2) = 1/(a'+2), \qquad a, a' \neq -2,$$
$$(ii) \qquad \eta = 2/(a'+2) = 4/(a+2), \qquad a, a' \neq -2.$$

Let $\hat{D}(b,a)$ be such an operator that $\hat{D}(b,a)W_a = W_b$ under the change of variable (141). Since (141) with option (i) or (ii) is invertible, the operator has an inverse. Hence $W_a = \hat{D}^{-1}(b,a)W_b$ in addition to $W_a = W_b$. Although the strict reciprocity is broken, we can talk about the power-dual symmetry in this relaxed sense.

Option (i): Transformation $r = C\rho^{2/(a+2)}$ in (141) brings

$$E_b - \Phi_b^2 = E_b - \lambda_b \rho^b - \lambda_{b'} \rho^{-1} - \frac{\hbar^2 \mu_b^2}{2m\rho^2}. \tag{148}$$

which contains a Coulomb-like potential in addition to a power potential for any value of a other than $a = -2$ ($a' = -2$). Option (i) must be associated with the substitutions,

$$E_b = -\eta^2 C^{2+a} \lambda_a, \quad \lambda_b = -\eta^2 C^2 E_a, \quad \lambda_{b'} = \eta^2 C^{2+a'} \lambda_{a'}, \quad \mu_b = \eta \mu_a, \tag{149}$$

and

$$\eta = \frac{2}{a+2} = \frac{1}{a'+2}, \quad b = -\frac{2a}{a+2}, \quad b' = -1. \tag{150}$$

The second relation in (149) may be used to determine the constant C of the transformation (141).

Option (ii): Transformation $r = C\rho^{2/(a'+2)}$ yields

$$E_b - \Phi_b^2 = E_b - \lambda_b \rho^2 - \lambda_{b'} \rho^{b'} - \frac{\hbar^2 \mu_b^2}{2m\rho^2}, \tag{151}$$

where a Hooke potential appears in addition to a power potential for any $a \neq -2$. Option (ii) comes with

$$E_b = -\eta^2 C^{2+a'} \lambda_{a'}, \quad \lambda_{b'} = -\eta^2 C^2 E_a, \quad \lambda_b = \eta^2 C^{2+a} \lambda_a, \quad \mu_b = \eta \mu_a, \tag{152}$$

and

$$\eta = \frac{2}{a'+2} = \frac{4}{a+2}, \quad b = 2, \quad b' = -\frac{2a'}{a'+2} = -\frac{2(a-2)}{a+2}. \tag{153}$$

Again, the second relation of (152) is able to fix the constant C.

Example 1. *The Coulomb–Hooke duality:*
Option (i) is appropriate for the Hooke to Coulomb transition with $a = 2$, $a' = 0$, $b = -1$ and $b' = -1$. By $r = C\rho^{1/2}$,

$$E_a - \Phi_a^2 = (E_a + \hbar\omega\mu_a) - \frac{1}{2}m\omega^2 r^2 - \frac{\hbar^2 \mu_a^2}{2mr^2}. \tag{154}$$

transforms to

$$E_b - \Phi_b^2 = \left(E_b - \frac{me^4}{2\hbar^2 \mu_b^2} \right) + \frac{e^2}{\rho} - \frac{\hbar^2 \mu_b^2}{2m\rho^2}, \tag{155}$$

where

$$E_b - \frac{me^4}{2\hbar^2 \mu_b^2} = -\frac{1}{8}m\omega^2 C^4, \quad C^2 = \frac{4e^2}{E_a + \hbar\omega\mu_a}, \quad \mu_b = \frac{1}{2}\mu_a. \tag{156}$$

Combining the first and the second relation of (156) gives

$$E_b = -\frac{2me^4}{\hbar^2 (E_a/\hbar\omega + \mu_a)^2} + \frac{me^4}{2\hbar^2 \mu_b^2}. \tag{157}$$

which can be converted to the QM spectrum for the hydrogen atom

$$E_b^{QM}(v, \ell) = E_b - \frac{me^4}{2\hbar^2 \mu_b^2} = -\frac{me^4}{2\hbar^2 (v + \ell + D/2 - 1/2)^2}, \tag{158}$$

by substitution of $E_a = 2\hbar\omega v$ and $\mu_a = 2\mu_b = 2(\ell + D/2 - 1/2)$.

Option (ii) is for the Coulomb to Hooke transition with $a = 0$, $a' = -1$, $b = 2$ and $b' = 2$. By $\rho = C^{-1} r^2$, the Equation (155) for the hydrogen atom transforms back to the Equation (154) for the radial oscillator. The constant C^{-1} appearing in the variable transformation is the inverse of C obtainable from the second relation of (156). Obviously, for the Coulomb–Hooke pair, option (ii) is the inverse of option (i). This confirms that the Coulomb–Hooke dual symmetry is valid in the SUSY semiclassical action.

Example 2. *A confinement problem:*

Option (i) and option (ii) may be used to study a confinement potential for which the superpotential (125) is of the form,

$$\Phi_a(r) = \epsilon \sqrt{\lambda_a} r^{1/2} - \frac{\hbar}{\sqrt{2m}} \frac{\mu_a}{r}, \qquad (\epsilon = 1, \ \lambda_a > 0). \tag{159}$$

Correspondingly, we have

$$E_a - \Phi_a^2(r) = E_a - \lambda_a r + \epsilon \hbar \mu_a \sqrt{\frac{2\lambda_a}{m}} r^{-1/2} - \frac{\hbar^2 \mu_a^2}{2mr^2}. \tag{160}$$

Option (i) with $a = 1$ $(a' = -1/2)$ gives $\eta = 2/3$. By $r = C\rho^{2/3}$, (159) transforms to

$$E_b - \Phi_b^2(\rho) = E_b - \lambda_b \rho^{-2/3} - \lambda_{b'} \rho^{-1} - \frac{\hbar^2 \mu_b^2}{2mr^2}, \tag{161}$$

where

$$E_b = -\frac{4}{9} C^3 \lambda_a, \ \lambda_b = -\frac{4}{9} C^2 E_a, \ \lambda_{b'} = -\epsilon \frac{4}{9} \hbar \mu_b C^{3/2} \sqrt{\frac{2\lambda_a}{m}}, \ \mu_b = \frac{2}{3} \mu_a. \tag{162}$$

The result (161) is not particularly interesting because it is not integrable. However, it is interesting that the limit $\lambda_b \to 0$ implies $E_a \to 0$. Hence the states in the vicinity of the zero-energy state of system A may be approximated by a set of states of the hydrogen atom.

Option (ii) with $a' = -1/2$ implies $\eta = 4/3$. The transformation $r = C\rho^{4/3}$ reduces $E_a - \Phi_a^2(r)$ of (160) to the form,

$$E_b - \Phi_b^2 = E_b - \lambda_b \rho^{2/3} - \lambda_{b'} \rho^2 - \frac{\mu_b^2 \hbar^2}{2m\rho^2}, \tag{163}$$

where

$$E_b = \epsilon \frac{16}{9} \hbar \mu_a C^{3/2} \sqrt{\frac{2\lambda_a}{m}}, \ \lambda_b = -\frac{16}{9} C^2 E_a, \ \lambda_{b'} = \frac{16}{9} C^3 \lambda_a, \ \mu_b = \frac{4}{3} \mu_a. \tag{164}$$

In the limit $\lambda_b \to 0$, system B becomes a radial harmonic oscillator with the coupling constant, $\lambda_{b'} > 0$. Thus, the states of system A in the vicinity of $E_a = 0$ may be approximated by those of such a radial harmonic oscillator. The confinement problem will be revisited in Section 2.4.

Remark 18. *The duality relation between system A and system B is reciprocal in the sense that the two systems are bijectively mapped to each other. Hence, if system A is dual to system B then system B is dual to system A. For instance, the Newton–Hooke duality in classical mechanics is reciprocal. The Newton–Hooke duality is the Hooke–Newton duality. The map from the Newton system to the Hooke system is bijective. By contrast, it has been known [63–65] that all the states of the hydrogen atom in three dimensions correspond to half the states of the isotropic harmonic oscillator in four dimensions. The map from the three dimensional Coulomb system (of $\ell_{cou} = 1, 2, 3, \dots$) to the four dimensional oscillator (of $\ell_{osc} = 2, 4, 6, \dots$) is injective. Hence all the states of the oscillator as a Hooke system (with $\ell_{osc} = 0, 1, 2, \dots$) cannot be mapped back to the Coulomb system (with $\ell_{cou} = 0, 1, 2, \dots$). The relation between the Coulomb system and the Hooke system at the level of the quantum structures is not reciprocal [64,66].*

Remark 19. *The Langer replacement, $\sqrt{\ell(\ell + D - 2)} \to \ell + (D - 2)/2$, is an ad hoc procedure introduced so as to be consistent with the quantum mechanical results [67]. In the literature [36], it has been suggested to regard the angular momentum L appearing in the Schrödinger equation as a continuous parameter since an arbitrary inverse square potential can be added to make the quantized angular momentum continuous. This reasoning, however, would make Langer's replacement nonsensical.*

Remark 20. *Recall that $\eta = -b/a$ for a dual pair (a, b) and that $\ell_b = \eta \ell_a$ and $D_b - 2 = \eta(D_a - 2)$. Although η can be any positive real number, in the following, we list a few examples of relevant numbers and relations for integral values of η:*

η	(a,b)	$\ell_a = 0,1,2,\dots$	$\ell_b = 0,1,2,\dots$	$D_a = 2,3,\dots$
2	$(-1,2)$	$\ell_b = 0,2,4,\dots$	$\ell_a = 0,1/2,1,\dots$	$D_b = 2D_a - 2$
3	$(-4/3,4)$	$\ell_b = 0,3,6,\dots$	$\ell_a = 0,1/3,2/3,\dots$	$D_b = 3D_a - 4$
4	$(-3/2,6)$	$\ell_b = 0,4,8,\dots$	$\ell_a = 0,1/4,1/2,\dots$	$D_b = 4D_a - 6$

For example, from the line of $\eta = 2$, we see that the states of the Coulomb system in $D_a = 3$ correspond to half the states of the Hooke system in $D_b = 4$. System A and system B cannot be reciprocal as long as the equality $\ell_b = \eta \ell_a$ is assumed.

Remark 21. *The time transformation $\mathfrak{T}$ has no role to play because the semiclassical action does not explicitly depend on time as a solution of the stationary Hamilton–Jacobi equation.*

Remark 22. *The condition $L^2 = 0$ assumed for the example in (117), if the Langer replacement (86) is employed, implies $\ell = 0$, which occurs only in two dimensions.*

Remark 23. *The spectrum (120) for $a = -1/2$ is similar to the approximate result obtained from an exact solution of Schrödinger's equation in one dimension [68].*

Remark 24. *The action on either side of (96) is not always integrable in closed form. Suppose the power a of the potential V_a be a non-zero integer. Then there are a few integrable examples. If $a = 2, -1$ or -2 then the action of system A is reducible to an elementary function, and if $a = 6, 4, 1, -3, -4$ or -6 then it can be expressed in terms of an elliptic function. Therefore, $(2, -1)$, $(-3, -6)$, $(-4, -4)$, $(1, -2/3)$, $(4, -1/3)$ and $(6, -3/2)$ are integrable dual pairs (a, b) when a is an integer other than 0 and -2 though b is not necessarily an integer. To $a = -2$, there corresponds the self-dual pair $(-2, -2)$ with $\eta = 1$.*

4. Power-Duality in Quantum Mechanics

The main object to be studied for the power-duality in quantum mechanics is the energy eigenequation of the form $\hat{H}|\psi\rangle = E|\psi\rangle$ where $\hat{H}$ is the Hamiltonian operator for a system in a power-law potential. Since one of the key operations in the power-duality transformation is the change of variable $r = C\rho^\eta$, we have to deal with the eigenequation in the radial coordinate representation, that is, the radial Schrödinger equation. In the context of the duality argument, the radial Schrödinger equation with power-law potentials have been exhaustively explored in the literature [36,39,40]. There is little room available to add something new. The aim of this section is, however, to present from the symmetry point of view the power-duality of the radial Schrödinger equation in parallel to the classical and semiclassical approaches. The power-duality in the path integral formulation of quantum mechanics is important but is not included in the present paper.

4.1. The Action for the Radial Schrödinger Equation

The stationary Schrödinger equation for a D dimensional system in a central-force potential $V(r)$ can be separated in polar coordinates into a radial equation and an angular part. The radial Schrödinger equation has the form,

$$\left\{ -\frac{\hbar^2}{2m}\left(\frac{d^2}{dr^2} + \frac{D-1}{r}\frac{d}{dr} \right) + \frac{\hbar^2 \ell(\ell + D - 2)}{2mr^2} + V(r) - E \right\} R_\ell(r) = 0. \tag{165}$$

In the above equation, the angular contribution appears in the third term, which stems from $\hat{L}^2 \mathcal{Y}_\ell^m(\mathbf{r}/r) = \ell(\ell + D - 2)\mathcal{Y}_\ell^m(\mathbf{r}/r)$ where $\hat{L}$ is the angular momentum operator and $\mathcal{Y}_\ell^m(\mathbf{r}/r)$ is the hyperspherical harmonics. Substituting $R_\ell(r) = r^{(1-D)/2}\psi_\ell(r)$ reduces it to a simplified differential equation on the positive half-line,

$$\left\{ -\frac{\hbar^2}{2m}\frac{d^2}{dr^2} + \frac{\hbar^2(L^2 - 1/4)}{2mr^2} + V(r) - E \right\} \psi_\ell(r) = 0, \tag{166}$$

where

$$L = \ell + (D - 2)/2, \qquad \ell = 0, 1, 2, \ldots. \tag{167}$$

For the sake of simplicity, we shall call Equation (166) the radial equation and $\psi_\ell(r)$ the wave function. The angular quantity L in (167) is precisely the same as Langer's choice (86) in the semiclassical action (see Remark 25). Under operation $\mathcal{L}: L_a = L_b/\eta$, the same problem that we have encountered in the semiclassical case should recur with the equality (167). Therefore, again, we adopt the view that the power-duality is basically a classical notion and follow the steps taken previously to circumvent the problem. Namely, for the duality argument, we treat L and E in (166) as continuous parameters. Only after the duality is established, we replace the parameters by their quantized counterparts. We consider that operation $\mathcal{L}$ applies only to the angular parameter and that $L_a = L_b/\eta$ does not imply $\ell_a + D/2 - 1 = (\ell_b + D/2 - 1)/\eta$. The last equality breaks the reciprocity that $\ell_a \in \mathbb{N}_0$ and $\ell_b \in \mathbb{N}_0$. The relation (167) holds true for each quantum system as an internal structure being independent of duality operations.

Suppose that system A has a two-term power potential $V_a(r) = \lambda_a r^a + \lambda_{a'} r^{a'}$ where $a \neq a'$. Defining the modified potential,

$$U_a(r) = \lambda_a r^a + \lambda_{a'} r^{a'} - E_a, \tag{168}$$

we write the radial Equation (166) as

$$\left\{ \frac{d^2}{dr^2} - \frac{L_a^2 - 1/4}{r^2} - \frac{2m}{\hbar^2}U_a(r) \right\} \psi_a(r) = 0. \tag{169}$$

Since we ignore the relation (167) for a while, we have dropped the subscript ℓ of the state function $\psi_a(r)$. The radial Equation (169) for system A is derivable from the following action integral,

$$W_a = \int_{\sigma_a} dr\, \mathcal{L}_a\left(\tfrac{d\psi_a^*(r)}{dr}, \tfrac{d\psi_a(r)}{dr}; \psi_a^*(r), \psi_a(r) \right), \tag{170}$$

having a fixed range $\sigma_a \ni r$ and the Lagrangian of the form,

$$\begin{aligned}
\mathcal{L}_a &= \frac{d\psi_a^*(r)}{dr}\frac{d\psi_a(r)}{dr} + \left(\frac{L_a^2 - 1/4}{r^2} + \frac{2m}{\hbar^2}U_a(r) \right)\psi_a^*(r)\psi_a(r) \\
&\quad - \frac{1}{2}\frac{d}{dr}\left(\psi_a^*(r)\frac{d\psi_a(r)}{dr} + \psi_a(r)\frac{d\psi_a^*(r)}{dr} \right),
\end{aligned} \tag{171}$$

where $\psi_a^*(r)$ is the complex conjugate of $\psi_a(r)$. Here we assume that the wave function $\psi_a(r)$ and its derivative are finite over the integration range σ_a. The last term of (171) is completely integrable, so that it contributes to the action as an unimportant additive constant. Use of the equality,

$$\frac{\mathrm{d}\psi_a^*(r)}{\mathrm{d}r}\frac{\mathrm{d}\psi_a(r)}{\mathrm{d}r} = -\psi_a^*(r)\frac{\mathrm{d}^2\psi_a(r)}{\mathrm{d}r^2} + \frac{\mathrm{d}}{\mathrm{d}r}\left(\psi_a^*(r)\frac{\mathrm{d}\psi_a(r)}{\mathrm{d}r}\right),\tag{172}$$

enables us to put the Lagrangian (171) into an alternative form,

$$\begin{aligned}\mathcal{L}_a' &= -\psi_a^*(r)\left\{\frac{\mathrm{d}^2\psi_a(r)}{\mathrm{d}r^2} - \left(\frac{L_a^2 - 1/4}{r^2} + \frac{2m}{\hbar^2}U_a(r)\right)\psi_a(r)\right\}\\ &\quad + \frac{1}{2}\frac{\mathrm{d}}{\mathrm{d}r}\left(\psi_a^*(r)\frac{\mathrm{d}\psi_a(r)}{\mathrm{d}r} - \psi_a(r)\frac{\mathrm{d}\psi_a^*(r)}{\mathrm{d}r}\right).\end{aligned}\tag{173}$$

The Euler–Lagrange equation, resulted from $\delta W/\delta\psi_a^* = 0$,

$$\frac{\mathrm{d}}{\mathrm{d}r}\left\{\frac{\partial\mathcal{L}_a}{\partial\left(\frac{\mathrm{d}\psi_a^*}{\mathrm{d}r}\right)}\right\} - \frac{\partial\mathcal{L}_a}{\partial\psi_a^*} = 0,\tag{174}$$

readily yields, with either of $\mathcal{L}_a$ or $\mathcal{L}_a'$, the radial Equation (169). Since $\mathcal{L}_a$ is symmetric with respect to $\psi(r)$ and $\psi^*(r)$, the complex conjugate of (169) can be derived from it. However, $\mathcal{L}_a'$ is inappropriate for deriving the radial equation for $\psi_a^*(r)$. For now we put $\mathcal{L}_a'$ aside even though there is no need for complex conjugation of the radial equation. For studying the power-duality in quantum mechanics, we focus our attention on the action W_a of (170) with the Lagrangian (171) rather than the radial Equation (169).

The symmetry operations that we consider for the power-duality in quantum mechanics are as follows

$$\mathfrak{R}:\ r = f(\rho) = C\rho^\eta\quad (C > 0),\tag{175}$$

$$\mathfrak{L}:\ L_b = \eta L_a,\tag{176}$$

$$\mathfrak{E}:\ E_b = -\eta^2 C^{a+2}\lambda_a,\quad \lambda_b = -\eta^2 C^2 E_a,\tag{177}$$

$$\mathfrak{C}:\ \eta = 2/(a+2) = (b+2)/2,\quad (a \neq -2,\ b \neq -2),\tag{178}$$

$$\mathfrak{B}:\ \lambda_{b'} = \lambda_{a'}(2/(a+2))^2 C^{a'+2},\quad b' = 2(a'-a)/(a+2),\tag{179}$$

$$\mathfrak{F}:\ \psi_a(r) = h(\rho)\psi_b(\rho).\tag{180}$$

In (180), $h(\rho)$ is a continuous positive real function of ρ.

As $\mathrm{d}r$ goes to $\mathrm{d}\rho$, the integration range of (170) changes from $\sigma_a \ni r$ to $\sigma_b \ni \rho$. Under (175) and (180), the first term of the Lagrangian (171) transforms as

$$\frac{\mathrm{d}\psi_a^*(r)}{\mathrm{d}r}\frac{\mathrm{d}\psi_a(r)}{\mathrm{d}r} = \frac{h^2}{f'^2}\left\{\frac{\mathrm{d}\psi_b^*(\rho)}{\mathrm{d}\rho}\frac{\mathrm{d}\psi_b(\rho)}{\mathrm{d}\rho} - \left[\frac{\mathrm{d}}{\mathrm{d}\rho}\left(\frac{h'}{h}\right) - \left(\frac{h'}{h}\right)^2\right]\psi_b^*\psi_b\right\} + \frac{h^2}{f'^2}\frac{\mathrm{d}}{\mathrm{d}\rho}\left(\frac{h'}{h}\psi_b^*\psi_b\right).\tag{181}$$

By choice, we let $h^2(\rho) = f'(\rho)$. Then the second term on the right-hand side of (181) reduces to the Schwarz derivative

$$\mathcal{S}[f] = \frac{f'''}{f'} - \frac{3}{2}\left(\frac{f''}{f'}\right)^2\tag{182}$$

divided by $2f'$. The third term of (181) can be decomposed to two terms by using the relation,

$$\frac{\mathrm{d}}{\mathrm{d}r}(\psi_a^*(r)\psi_a(r)) = \frac{h^2}{f'}\frac{\mathrm{d}}{\mathrm{d}\rho}(\psi_b^*(\rho)\psi_b(\rho)) + \frac{2hh'}{f'}\psi_b^*\psi_b.\tag{183}$$

Therefore,

$$\frac{\mathrm{d}\psi_a^*(r)}{\mathrm{d}r}\frac{\mathrm{d}\psi_a(r)}{\mathrm{d}r} - \frac{1}{2}\frac{\mathrm{d}}{\mathrm{d}r}\left[\frac{\mathrm{d}}{\mathrm{d}r}(\psi_a^*(r)\psi_a(r))\right] = \frac{1}{f'}\left\{\frac{\mathrm{d}\psi_b^*(\rho)}{\mathrm{d}\rho}\frac{\mathrm{d}\psi_b(\rho)}{\mathrm{d}\rho} - \frac{1}{2}\mathcal{S}[f]\psi_b^*\psi_b\right\} - \frac{1}{2f'}\frac{\mathrm{d}}{\mathrm{d}\rho}\left[\frac{\mathrm{d}}{\mathrm{d}\rho}(\psi_b^*(\rho)\psi_b(\rho))\right]. \tag{184}$$

The angular term of the Lagrangian (171) transforms as

$$\frac{L_a^2 - 1/4}{r^2}\psi_a^*(r)\psi_a(r) = \frac{1}{f'}\frac{g(L_a^2 - 1/4)}{f^2}\psi_b^*(\rho)\psi_b(\rho) \tag{185}$$

where g denotes f'^2 as in the classical and semiclassical cases. The energy-potential term of (171) changes as

$$\frac{2m}{\hbar^2}U_a(r)\psi_a^*(r)\psi_a(r) = \frac{2m}{\hbar^2}gU_a(f(\rho))\psi_b^*(\rho)\psi_b(\rho). \tag{186}$$

Moreover, we let $f(\rho) = C\rho^\eta$ as defined by (175). Then $\mathcal{S}[f] = -(\eta^2 - 1)/2$, $g = C^2\eta^2\rho^{2\eta-2}$ and $g/f^2 = C^2\eta^2\rho^2$. Hence, we have

$$g(L_a^2 - 1/4)/f^2 - (1/2)\mathcal{S}[f] = (\eta^2 L_a^2 - 1/4)/\rho^2, \tag{187}$$

which results in $(L_b^2 - 1/4)/\rho^2$ under $\mathfrak{L}: L_b = \eta L_a$. Changing the variable by (175) and making the energy-coupling exchange by (177) result in

$$g(\rho)U_a(C\rho^\eta) = -E_b\rho^{a\eta+2\eta-2} + C^{a'+2}\lambda_{b'}\rho^{a'\eta+2\eta-2} + \lambda_b\rho^{2\eta-2}, \tag{188}$$

which is written as

$$U_b(\rho) = \lambda_b\rho^b + \lambda_{b'}\rho^{b'} - E_b \tag{189}$$

with the help of (178) and (179). Namely, $U_a(r)$ goes to $U_b(\rho)$ by $U_b(\rho) = g(\rho)U_a(r)$. Consequently, we obtain $W_a = W_b$ or, emphasizing the parameter dependence of the Lagrangian,

$$\int_{\sigma_a} \mathrm{d}r\, \mathcal{L}_a(\lambda_a, L_a, U_a) = \int_{\sigma_b} \mathrm{d}\rho\, \mathcal{L}_b(\lambda_b, L_b, U_b), \tag{190}$$

where

$$\mathcal{L}_b = \frac{\mathrm{d}\psi_b^*(\rho)}{\mathrm{d}\rho}\frac{\mathrm{d}\psi_b(\rho)}{\mathrm{d}\rho} + \left(\frac{L_b^2 - 1/4}{\rho^2} + \frac{2m}{\hbar^2}U_b(\rho)\right)\psi_b^*(\rho)\psi_b(\rho) - \frac{1}{2}\frac{\mathrm{d}}{\mathrm{d}\rho}\left(\psi_b^*(\rho)\frac{\mathrm{d}\psi_b(\rho)}{\mathrm{d}\rho} + \psi_b(\rho)\frac{\mathrm{d}\psi_b^*(\rho)}{\mathrm{d}\rho}\right). \tag{191}$$

The last term of (191) is completely integrable and contributes to W_b as an unimportant constant. We identify $\mathcal{L}_b$ of (191) with the Lagrangian of system B, use of which leads to the radial equation for system B,

$$\left\{\frac{\mathrm{d}^2}{\mathrm{d}\rho^2} - \frac{L_b^2 - 1/4}{\rho^2} - \frac{2m}{\hbar^2}U_b(\rho)\right\}\psi_b(\rho) = 0. \tag{192}$$

Apparently the form of the Lagrangian is preserved under the set of power-duality operations, $\{\mathfrak{R}, \mathfrak{L}, \mathfrak{C}, \mathfrak{E}, \mathfrak{B}, \mathfrak{F}\}$. Furthermore, with the Lagrangians $\mathcal{L}_a$ of (171) and $\mathcal{L}_b$ of (191), the equality (190) implies that the action W of (170) is invariant under the same set of operations. By (190) the complex conjugate of the radial Schrödinger Equation (166) is as well assured to be form-invariant.

To complete the procedure, as we have done for the semiclassical case, we must replace in an ad hoc manner each of the angular momentum parameters by the quantized form $\ell + (D-2)/2$ with $\ell = 0, 1, 2, \ldots$. Using the dot-equality introduced in Section 3.1, we write the form-invariance of the action amended by the angular quantization with $\ell_a, \ell_b \in \mathbb{N}_0$,

$$\int_{\sigma_a} dr\, \mathcal{L}_a(\lambda_a, \ell_a + (D-2)/2, U_a) \doteq \int_{\sigma_b} d\rho\, \mathcal{L}_b(\lambda_b, \ell_b + (D-2)/2, U_b), \tag{193}$$

which warrants that the radial Schrödinger Equation (166) with the angular quantization (167) is form-invariant under the set of duality operations, $\{\mathfrak{R}, \mathfrak{L}, \mathfrak{C}, \mathfrak{E}, \mathfrak{B}, \mathfrak{F}\}$. In this modified sense we claim that two quantum systems with $V_a(r) = \lambda_a r^a + \lambda_{a'} r^{a'}$ and with $V_b(\rho) = \lambda_b \rho^b + \lambda_{b'} \rho^{b'}$ are in power-duality provided that $(a+2)(b+2) = 4$.

4.2. Energy Formulas, Wave Functions and Green Functions

In arriving at the invariance relation (190), we have seen the equality $dr\, \mathcal{L}_a = d\rho\, \mathcal{L}_b$ under the duality operations. The relation (190) is valid with the alternative Lagrangian $\mathcal{L}'$ of (173), suggesting $dr\, \mathcal{L}'_a = d\rho\, \mathcal{L}'_b$. The last equality in turn leads to

$$\left\{ \frac{d^2}{dr^2} - \frac{L_a^2 - 1/4}{r^2} - \frac{2m}{\hbar^2} U_a(r) \right\} \psi_a(r) = \frac{1}{h^3} \left\{ \frac{d^2}{d\rho^2} - \frac{L_b^2 - 1/4}{\rho^2} - \frac{2m}{\hbar^2} U_b(\rho) \right\} \frac{1}{h} \psi_a(f(\rho)) \tag{194}$$

where $f' = h^2 = C\eta\rho^{\eta-1}$. Let $H_a(r)$ be the Hamiltonian for system A in the r-representation, that is,

$$H_a(r) = -\frac{\hbar^2}{2m} \frac{d^2}{dr^2} + \frac{\hbar^2(L_a^2 - 1/4)}{2mr^2} + \lambda_a r^a + \lambda_{a'} r^{a'}. \tag{195}$$

Similarly, we define $H_b(\rho)$ for system B. By using the exchange symbol $X(b,a)$, we have $H_b(\xi_b) - E_b = X(b,a)\{H_a(\xi_a) - E_a\}$ where $\xi_a = r$ and $\xi_b = \rho$. Then the equality (194) may be put into the form,

$$\{H_a(r) - E_a\}\psi_a(r) = \frac{1}{h^3}\{H_b(\rho) - E_b\}\psi_b(\rho), \tag{196}$$

when $\psi_a(r) = h(\rho)\psi_b(\rho)$ with $r = f(\rho)$ and $f' = h^2$. Evidently, the radial Equation (169), expressed as $\{H_a(r) - E_a\}\psi_a(r) = 0$, implies $\{H_b(\rho) - E_b\}\psi_b(\rho) = 0$.

4.2.1. Energy Formulas

To find the energy spectrum of system A, we usually solve the radial Equation of (169) by specifying boundary conditions on $\psi_a(r)$. Suppose we found a solution $\psi_a(r; \nu)$ compatible with the given boundary conditions when the energy parameter took a specific value $E_a(\nu)$ characterized by a real number ν. This solution may be seen as an eigenfunction satisfying

$$H_a(r)\psi_a(r; \nu) = E_a(\nu)\psi_a(r; \nu). \tag{197}$$

Since operation $\mathfrak{F}$ demands $\psi_a(r; \nu) = \psi_a(f(\rho); \nu) = h(\rho)\psi_b(\rho; \nu)$, the Equation (197) should imply via the equality (196)

$$H_b(\rho)\psi_b(\rho; \nu)) = E_b(\nu)\psi_b(\rho; \nu). \tag{198}$$

This shows that the number ν is a dual invariant being common to $E_a(\nu)$ and $E_b(\nu)$. As has been repeatedly mentioned earlier, the duality operations cannot interfere the internal structure of each quantum system. In general, there are a number of solutions for the given boundary conditions. Thus, ν may be representing a set of numbers. Then we understand that the value of ν is preserved by $\mathfrak{F}$. For a while, however, we treat ν as another parameter and express the energy E_a as a function of λ_a, L_a and ν,

$$E_a = E_a(\lambda_a, L_a, \nu). \tag{199}$$

This corresponds to the energy function $E_a(\lambda_a, L_a, N)$ in the semiclassical case. We convert this energy function to the energy spectrum of system A by replacing the parameters L_a and v to their quantum counterparts. If we restrict our interest to bound state solutions, the parameter v is to be replaced by a set of discrete numbers $v = 0, 1, 2, \ldots$. Furthermore, putting the angular parameter L_a into the Langer form (167), we obtain the discrete energy spectrum of system A,

$$E_a(\ell_a, v) = E_a(\lambda_a, \ell_a + D/2 - 1, v), \tag{200}$$

where $\ell_a \in \mathbb{N}_0$ and $v \in \mathbb{N}_0$.

Since the energy functions $E_a(\lambda_a, L_a, v)$ and $E_b(\lambda_b, L_b, v)$ are related by the classical energy formulas, (60) and (64)–(65), the corresponding energy spectra $E_a(\ell_a, v)$ and $E_b(\ell_b, v)$ can be related by the same formulas provided the angular parameter and the quantum parameter are properly expressed in terms of quantum numbers. Knowing the energy spectrum of the form $E_a(\ell_a, v) = \mathcal{E}(\lambda_a, L_a, v)$ for system A, we can determine the energy spectrum E_b of system B by

$$E_b(\ell_b, v) = -\eta^2 C^{a+2} \mathcal{E}^{-1}(-\lambda_b/(\eta^2 C^2), L_b/\eta, v), \tag{201}$$

where $L_b = \ell_b + D/2 - 1$ with $\ell_b \in \mathbb{N}_0$. For the bound state spectrum, $v = 0, 1, 2, \ldots$.

If the energy spectrum of system A is given in the form

$$E_a(\ell_a, v) = \pm \frac{1}{4}(a+2)^2 |\lambda_a|^{2/(a+2)} \left[\mathcal{F}\left(\sqrt{2/(a+2)}\, (\ell_a + D/2 - 1), v \right) \right]^{1/a} \tag{202}$$

then the energy spectrum of system B is given by

$$E_b(\ell_b, v) = \pm \frac{1}{4}(b+2)^2 |\lambda_b|^{2/(b+2)} \left[\mathcal{F}\left(\sqrt{2/(b+2)}\, (\ell_b + D/2 - 1), v \right) \right]^{1/b}. \tag{203}$$

These relations are the same as the semiclassical relations (104 and (105) where the signs are determined by the signs of the coupling constants, $\mathrm{sgn}\, E_a = -\mathrm{sgn}\, \lambda_b$ and $\mathrm{sgn}\, E_b = -\mathrm{sgn}\, \lambda_a$.

4.2.2. Wave Functions

The wave function transforms as $\psi_a(r; L_a, v) = h(\rho)\psi_b(\rho; L_b, v)$. Therefore, if an eigenfunction of system A is given, then the corresponding eigenfunction of system B can be determined by

$$\psi_b(\rho; L_b, v) = \frac{1}{h(\rho)} \psi_a(C\rho^\eta; L_b/\eta, v), \tag{204}$$

where $L_b = \ell_b + D/2 - 1$ with $\ell_b \in \mathbb{N}_0$. Both $\psi_a(r)$ and $\psi_b(\rho)$ as eigenfunctions are supposed to be square-integrable, and each of them must be normalizable to unity. However, even if $\psi_a(r)$ is normalized to unity, it is unlikely that $\psi_b(\rho)$ constructed by (204) is normalized to unity. This is because

$$\int_0^\infty dr\, |\psi_a(r)|^2 = \int_0^\infty d\rho\, g(\rho)\, |\psi_b(\rho)|^2 = 1 \tag{205}$$

where $g(\rho) = [f'(\rho)]^2 = [h(\rho)]^4 = C^2\eta^2\rho^{2(\eta-1)}$. In this regard, if system A and system B are power-dual to each other, the formula (204) determines $\psi_b(\rho)$ of system B out of $\psi_a(r)$ of system A except for the normalization.

4.2.3. Green Functions

The Green function $G(r, r'; z) = \langle r|\hat{G}(z)|r'\rangle$ is the r-representation of the resolvent $\hat{G}(z) = (z - \hat{H})^{-1}$ where $z \in \mathbb{C}\backslash\mathrm{spec}\,\hat{H}$ and $\hat{H}$ is the Hamiltonian operator of the system in question. Let $E(v)$ and $|\psi(v)\rangle$ be the eigenvalue of $\hat{H}$ and the corresponding eigenstate, respectively, so that $\hat{H}|\psi(v)\rangle = E(v)|\psi(v)\rangle$. For simplicity, we consider the case where $v \in \mathbb{N}_0$. Assume the eigenstates are orthonormalized and form a complete set, that is,

$$\langle \psi(v)|\psi(v')\rangle = \delta_{v,v'}, \qquad \sum_{v \in \mathbb{N}_0} |\psi(v)\rangle\langle\psi(v)| = 1. \tag{206}$$

From the completeness condition in (206), it is obvious that

$$\hat{G}(z) = \sum_{v \in \mathbb{N}_0} \frac{|\psi(v)\rangle\langle\psi(v)|}{z - E(v)}. \tag{207}$$

Hence, the Green function can be written as

$$G(r,r';z) = \sum_{v \in \mathbb{N}_0} \frac{\psi^*(r';v)\psi(r;v)}{z - E(v)}. \tag{208}$$

Use of Cauchy's integral formula leads us to the expression,

$$\psi^*(r,v)\psi(r';v) = \frac{1}{2\pi i}\oint_{C_v} dz\, G(r,r';z), \tag{209}$$

where the closed contour C_v counterclockwise encloses only the simple pole $z = E(v)$ for a fixed value of v. Note that we will deal only with radial, hence one-dimensional, problems where no degeneracies can occur. Multiplying both sides of (209) by two factors $v(r)$ and $v(r')$ yields

$$\tilde{\psi}^*(r,v)\tilde{\psi}(r';v) = \frac{1}{2\pi i}\oint_{C_v} dz\, \tilde{G}(r,r';z), \tag{210}$$

where $\tilde{\psi}(r;v) = v(r)\psi(r;v)$ and $\tilde{G}(r,r';z) = v(r)v(r')G(r,r';z)$.

For instance, the Green function $\mathcal{G}(r,r';E)$ for the radial Schrödinger Equation (165) is related to the Green function $G(r,r';E)$ for the simplified radial Equation (166) by

$$\mathcal{G}(r,r';E) = (r\,r')^{(1-D)/2}G(r,r';E) \tag{211}$$

as the wave functions of (165) and (166) are connected by $R_\ell(r) = r^{(1-D)/2}\psi_\ell(r)$.

Suppose the Green functions of system A and system B are given, respectively, by

$$\psi_a^*(r;v)\psi_a(r';v) = \frac{1}{2\pi i}\oint_{C_v} dz\, G_a(r,r';z), \tag{212}$$

and

$$\psi_b^*(\rho;v)\psi_b(\rho';v) = \frac{1}{2\pi i}\oint_{C_v} dz\, G_b(\rho,\rho';z). \tag{213}$$

By comparing these two expressions, we see that if $\psi_a(r;v) = h(\rho)\psi_b(\rho;v)$ then

$$G_a(r,r';E_a(v)) = h(\rho)h(\rho')G_b(\rho,\rho',E_b(v)). \tag{214}$$

The above result is obtained without considering the detail of the Hamiltonian. In the following, an alternative account is provided for deriving the same result by using the Hamiltonian explicitly. Let $\hat{H}_a$ be the Hamiltonian operator of system A such that $\langle r|\hat{H}_a - E_a|r'\rangle = (H_a(r) - E_a)\langle r|r'\rangle$. Then it is obvious that

$$\{H_a(r) - E_a\}G_a(r,r';E_a) = -\delta(r - r'). \tag{215}$$

According to (194), Equation (215) implies

$$\{H_b(\rho) - E_b\}\frac{1}{h}G_a(f(\rho),f(\rho');E_a) = -h^3(\rho)\delta(f(\rho) - f(\rho')). \tag{216}$$

From the relations,

$$\int dr\,|r\rangle\langle r| = \int d\rho\, f'(\rho)|f(\rho)\rangle\langle f(\rho')| = \int d\rho\, |\rho\rangle\langle \rho'| = 1, \tag{217}$$

there follows $|\rho\rangle = h(\rho)|f(\rho)\rangle$. Hence we have, $\langle \rho|\rho'\rangle = h(\rho)h(\rho')\langle f(\rho)|f(\rho')\rangle$, that is, $\delta(r-r') = \delta(f(\rho)-f(\rho')) = [h(\rho)h(\rho')]^{-1}\delta(\rho-\rho')$. Thus, we arrive at the radial equation satisfied by the Green function of system B,

$$\{H_b(\rho)-E_b\}G_b(\rho,\rho';E_b) = -\delta(\rho-\rho'), \tag{218}$$

if the Green function transforms as

$$G_a(r,r_0;E_a,L_a) = h(\rho)h(\rho')G_b(\rho,\rho';E_b,L_b). \tag{219}$$

Substitution of $L_a = \ell_a + D/2 - 1$ with $\ell_a \in \mathbb{N}_0$ and $L_b = \ell_b + D/2 - 1$ with $\ell_b \in \mathbb{N}_0$ into (219) results in

$$G_a(r,r';E_a,\ell_a+D/2-1) \doteq h(\rho)h(\rho_0)G_b(\rho,\rho';E_b,\ell_b+D/2-1), \tag{220}$$

which is not an equality as $\ell_a \in \mathbb{N}_0$ and $\ell_b \in \mathbb{N}_0$ are assumed. Insofar as system B is power-dual to system A, the Green function of system B can be expressed in terms of the Green function of system A as

$$G_b(\rho,\rho';E_b,\ell_b+D/2-1,\lambda_b,\lambda_{b'}) = [(f'(\rho)f'(\rho'))]^{-1/2}G_a\big(f(\rho),f(\rho');E_a,(\ell_b+D/2-1)/\eta,\lambda_a,\lambda_{a'}\big) \tag{221}$$

where $f(\rho) = C\rho^\eta$ and the parameters E_a, λ_a and $\lambda_{a'}$ are given via the relations (177) and (179) in terms of E_b, λ_b and $\lambda_{b'}$. This relation is an equality even though (220) is a dot equality. An expression similar to but slightly different from (221) has been obtained by Johnson [36] in much the same way.

4.3. The Coulomb–Hooke Dual Pair

Again, we take up the Coulomb–Hooke dual pair to test the transformation properties shown in Section 3.1. Let system A be the hydrogen atom with $\lambda_a = -e^2 < 0$ and system B a radial oscillator with $\lambda_b = \frac{1}{2}m\omega^2 > 0$. So $(a,b) = (-1,2)$ and $\eta = -b/a = 2$. Both systems are assumed to be in D dimensional space. The Coulomb system has the scattering states $(E_a > 0)$ as well as the bound states $(E_a < 0)$. However, the exchange relations (177) prohibits the process $(E_a > 0, \lambda_a < 0) \Rightarrow (E_b > 0, \lambda_b > 0)$. The Coulomb–Hooke duality occurs only when the Coulomb system is in bound states.

The energy relations: Suppose we know that the energy spectrum of system A has the form,

$$E_a(\lambda_a,L_a,\nu) = -\frac{me^4}{2\hbar^2(\nu+L_a+1/2)^2}, \tag{222}$$

where $\lambda_a = -e^2$, $\nu \in \mathbb{N}_0$ and $L_a = \ell_b + D/2 - 1$ with $\ell_a \in \mathbb{N}_0$. Then the formula (202) leads to

$$\mathcal{F}\left(\sqrt{2}L_a,\nu\right) = \frac{\hbar^2}{2m}\left[\nu+(\sqrt{2}L_a)/\sqrt{2}+1/2\right]^2. \tag{223}$$

Careful use of this result in the formula (203) enables us to determine the energy spectrum of system B. Namely,

$$E_b(\lambda_b,L_b,\nu) = 4\sqrt{\lambda_b}\sqrt{\frac{\hbar^2}{2m}}\left[\nu+(L_b/\sqrt{2})/\sqrt{2}+1/2\right]^{1/2}. \tag{224}$$

Substituting $\lambda_b = m\omega^2/2$ and $L_b = \ell_b + D/2 - 1$ in (224), we reach the standard expression for the energy spectrum of the isotropic harmonic oscillator in D-dimensional space,

$$E_b(\ell_b, \nu) = \hbar\omega(2\nu + \ell_b + D/2) \quad (\ell_b, \nu \in \mathbb{N}_0). \tag{225}$$

Wave functions: The radial Equation (166) for the Coulomb potential $V(r) = -e^2/r$ can easily be converted to the Whittaker Equation [69]

$$\left\{ \frac{d^2}{dx^2} - \frac{L^2 - 1/4}{x^2} + \frac{k}{x} - \frac{1}{4} \right\} w(x) = 0, \tag{226}$$

where $L = \ell + D/2 - 1$ ($\ell \in \mathbb{N}_0$). In the conversion, we have let $x = 2\kappa r$, $k = me^2/(\hbar^2\kappa) = k_a$, $\hbar\kappa = \sqrt{-2mE}$, $L = L_a$ and $w(x) = \psi_a(x/(2\kappa))$. This set of replacements is indeed a duality map for the self-dual pair $(a, a) = (-1, -1)$. The Whittaker functions, $M_{k,L}(x)$ and $W_{k,L}(x)$, are two linearly independent solutions of the Whittaker Equation (226). For $|x|$ small, $M_{k,L}(x) \sim x^{L+1/2}$ and $W_{k,L}(x) \sim -\frac{\Gamma(2L)}{\Gamma(L-k+1/2)} x^{-L+1/2}$. If $-\pi/2 < \arg x < 3\pi/2$ and $|x|$ is large, then

$$M_{k,L}(x) \sim \Gamma(2L+1) \left\{ \frac{e^{i\pi(L-k+\frac{1}{2})} e^{-x/2} x^k}{\Gamma(L+k+\frac{1}{2})} + \frac{e^{x/2} x^{-k}}{\Gamma(L-k+\frac{1}{2})} \right\}, \tag{227}$$

and, if $x \notin \mathbb{R}^-$ and $|x|$ is large,

$$W_{k,L}(x) \sim e^{-x/2} x^k [1 + O(x^{-1})]. \tag{228}$$

The first solution $M_{k,L}(x)$ vanishes at $x = 0$ as $L > -1/2$ but diverges as $|x| \to \infty$ unless $k - L - \frac{1}{2} \in \mathbb{N}_0$, whereas the second solution $W_{k,L}(x)$ diverges at $x = 0$ but converges to zero as $|x| \to \infty$.

The solution for the Coulomb problem is given in terms of the Whittaker function,

$$\psi_a(r; L_a, \nu) = \mathcal{N}_a(L_a) M_{\nu+L_a+\frac{1}{2}, L_a}(2\kappa r), \tag{229}$$

where k_a is replaced by $\nu + L_a + 1/2$. For the bound state solution which vanishes at infinity, we have to let $\nu = 0, 1, 2, \ldots$. In this case, $k_a = \nu + L_a + 1/2$ implies the discrete spectrum $E_a(\lambda_a, L_a, \nu)$ in (222).

Since the Whittaker function $M_{k,\mu}(z)$ is related to the Laguerre function $L_\nu^{2\mu}(z)$ as

$$M_{\mu+\nu+\frac{1}{2}, \mu}(z) = \frac{\Gamma(2\mu+1)\Gamma(\nu+1)}{\Gamma(2\mu+\nu+1)} e^{-z/2} z^{\mu+\frac{1}{2}} L_\nu^{2\mu}(z), \tag{230}$$

the eigenfunction may also be expressed in terms of the Laguerre function as

$$\psi_a(r; L_a, \nu) = \mathcal{N}_a(L_a) \frac{\Gamma(2L_a+1)\Gamma(\nu+1)}{\Gamma(\nu+2L_a+1)} e^{-\kappa r} (2\kappa r)^{L_a+\frac{1}{2}} L_\nu^{2L_a}(2\kappa r), \tag{231}$$

which is normalized to unity with

$$\mathcal{N}_a(L_a) = \frac{\hbar\kappa/\sqrt{me^2}}{\Gamma(2L_a+1)} \sqrt{\frac{\Gamma(\nu+2L_a+1)}{\Gamma(\nu+1)}}. \tag{232}$$

The radial equation for the Hooke system with $V_b(\rho) = \frac{1}{2}m\omega^2\rho^2$, too, can be reduced to the Whittaker equation by letting

$$y = (m\omega/\hbar)\rho^2, \quad L = L_b/2, \quad k = E_b/(2\hbar\omega) = k_b, \quad w(y) = y^{1/4}\psi_b(y), \tag{233}$$

which form a duality map for $(b,c) = (2,-1)$. Here $L_b = \ell_b + D/2 - 1$ with $\ell_b \in \mathbb{N}_0$. The bound state solution for the radial oscillator is given by

$$\psi_b(\rho; L_b, \nu) = \mathcal{N}_b(L_b) \frac{1}{\sqrt{\rho}} M_{\nu+\frac{1}{2}L_b+\frac{1}{2}, \frac{1}{2}L_b}\left(\frac{m\omega}{\hbar}\rho^2\right). \tag{234}$$

The choice $k_b = (2\nu + L_b + 1)/2$ with $\nu \in \mathbb{N}_0$ makes the solution (234) the eigenfunction belonging to the energy $E_b(\nu, \ell_b)$ in (224). In terms of the Laguerre function, it reads

$$\psi_b(\rho; L_b, \nu) = \mathcal{N}_b(L_b) \frac{\Gamma(L_b+1)\Gamma(\nu+1)}{\Gamma(\nu+L_b+1)} e^{-(m\omega/2\hbar)\rho^2}\left(\frac{m\omega}{\hbar}\rho^2\right)^{(L_b+\frac{1}{2})/2} L_\nu^{L_b}\left(\frac{m\omega}{\hbar}\rho^2\right), \tag{235}$$

which is normalized to unity with

$$\mathcal{N}_b(L_b) = \frac{(4m\omega/\hbar)^{1/4}}{\Gamma(L_b+1)} \sqrt{\frac{\Gamma(\nu+L_b+1)}{\Gamma(\nu+1)}}. \tag{236}$$

The process of going from (229) to (234) is rather straightforward. First we notice that $\eta = -b/a = 2$ for the Coulomb–Hooke pair $(a,b) = (-1,2)$. Then we use the relation $\lambda_b = -\eta^2 C^2 E_a$, $\lambda_b = m\omega^2/2$ and $\hbar\kappa = \sqrt{-2mE_a}$ to get $C = m\omega/(2\hbar\kappa)$. Hence operation $\mathfrak{R}: r = C\rho^\eta$ with $\eta = 2$ yields $2\kappa r = (m\omega/\hbar)\rho^2$. In addition, we apply $\mathfrak{L}: L_a = L_b/2$. Consequently, we have the right hand side of (204) for $a = -1, \eta = 2$ and $h(\rho) = \sqrt{m\omega/(\hbar\kappa)}\rho^{1/2}$ in the form,

$$\sqrt{\hbar\kappa/m\omega} \frac{1}{\sqrt{\rho}} \psi_a((m\omega/2\hbar\kappa)\rho^2; L_b/2, \nu) = \tilde{\mathcal{N}}_b(L_b) \frac{1}{\sqrt{\rho}} M_{\nu+\frac{1}{2}L_b+\frac{1}{2}, \frac{1}{2}L_b}\left(\frac{m\omega}{\hbar}\rho^2\right), \tag{237}$$

which coincides with the eigenfunction for the radial oscillator in (234) except for the normalization factor. In (237),

$$\tilde{\mathcal{N}}_b(L_b) = \sqrt{\hbar\kappa/m\omega}\,\mathcal{N}_a(L_b/2), \tag{238}$$

which differs from $\mathcal{N}_b(L_b)$ of (236) due to the difference of factors, $\sqrt{\hbar^2\kappa^3/(me^2)}(m\omega/\hbar)^{-1/2} \neq \sqrt{2}(m\omega/\hbar)^{1/4}$. The wave function of the radial oscillator can be determined by the radial wave function of the hydrogen atom except for its normalization.

The Green functions: The Green function of interest, $G_a(r,r'; E, L)$, obeys the radial equation,

$$\left\{\frac{d^2}{dr^2} - \frac{L^2-1/4}{r^2} - \frac{2m}{\hbar^2}V(r) + \frac{2m}{\hbar^2}E\right\}G_a(r,r'; E, L) = -\frac{2m}{\hbar^2}\delta(r-r'), \tag{239}$$

where $V_a(r) = \lambda_a r^a + \lambda_{a'} r^{a'}$. The boundary conditions we impose on it are

$$\lim_{r\to 0} G(r,r'; E, L) = 0 \quad \text{and} \quad \lim_{r\to\infty} G(r,r'; E, L) < \infty. \tag{240}$$

Let $\psi^{(1)}(r)$ and $\psi^{(2)}(r)$ be two independent solutions of the radial Equation (166). Let us assume that $\psi^{(1)}(r)$ remains finite as $r \to \infty$ while the second solution obeys $\psi^{(2)}(0) = 0$. With these solutions, following the standard procedure [70], we can construct the Green function $G(r,r'; E, L)$ as

$$G(r,r'; E, L) = \frac{2m}{\hbar^2 W[\psi^{(1)}, \psi^{(2)}]} \begin{cases} \psi^{(1)}(r)\,\psi^{(2)}(r'), & r > r' \\ \psi^{(1)}(r')\,\psi^{(2)}(r), & r' > r \end{cases} \tag{241}$$

where $W[\cdot, \cdot]$ signifies the Wronskian.

For the Coulomb problem with $V_a(r) = -e^2 r^{-1}$, we let $\psi^{(1)}(r) = W_{k_a,L_a}(2\kappa r)$ and $\psi^{(2)}(r) = M_{k_a,L_a}(2\kappa r)$. Then we calculate the Wronskian to get

$$(2\kappa)^{-1}\mathcal{W}[W_{k,L}(2\kappa r), M_{k,L}(2\kappa r)] = \mathcal{W}[W_{k,L}(x), M_{k,L}(x)] = -\frac{\Gamma(2L+1)}{\Gamma(L-k+\frac{1}{2})}, \qquad (242)$$

where we have use the property,

$$\mathcal{W}[W_{k,L}(x), M_{k,L}(x)] = (dy/dx)\mathcal{W}[W_{k,L}(y), M_{k,L}(y)]. \qquad (243)$$

Substituting this result in the formula (241), we obtain the radial Green function for the Coulomb problem,

$$G_a(r,r'; E_a, L_a) - -\frac{m}{\hbar^2\kappa}\frac{\Gamma(L_a - k_a + \frac{1}{2})}{\Gamma(2L_a+1)}W_{k_a,L_a}(2\kappa r_>)\, M_{k_a,L_a}(2\kappa r_<), \qquad (244)$$

where $r_> = \max\{r,r'\}$ and $r_< = \min\{r,r'\}$. We have also set $\kappa = \sqrt{-2mE_a}\,\hbar$ and $k_a = me^2/(\hbar\sqrt{-2mE_a})$, both of which are in general complex numbers. The resultant Green function is a double-valued function of E_a. It contains the contribution from the continuous states (corresponding to the branch-cut along the positive real line on E_a) as well as the bound states (corresponding to the poles on the negative real axis). The poles of $G(r,r'; L_a, E_a)$ on the E_a-plane occur when $L_a - k_a + \frac{1}{2} = -\nu$ with $\nu \in \mathbb{N}_0$, yielding the discrete energy spectrum (222).

Similarly, for the radial oscillator with $V_b(\rho) = (m/2)\omega^2\rho^2$, we let $\psi^{(1)}(\rho) = W_{k_b,L_b}((m\omega/\hbar)\rho^2)$ and $\psi^{(2)}(\rho) = M_{k_b,L_b}((m\omega/\hbar)\rho^2)$. Use of the property,

$$\mathcal{W}[\chi(y)W_{k,L}(y), \chi(y)M_{k,L}(y)] = [\chi(y)]^2 \mathcal{W}[W_{k,L}(y), M_{k,L}(y)], \qquad (245)$$

for a differentiable function $\chi(y)$, together with (243) and (242), enables us to evaluate the Wronskian and to get to the Green function for the radial oscillator,

$$G_b(\rho,\rho'; L_b, E_b) = -\frac{1}{\hbar\omega}\frac{1}{\sqrt{\rho\rho'}}\frac{\Gamma(\frac{1}{2}L_b - k_b + \frac{1}{2})}{\Gamma(L_b+1)}W_{k_b,\frac{1}{2}L_b}\left(\frac{m\omega}{\hbar}\rho_>^2\right) M_{k_b,\frac{1}{2}L_b}\left(\frac{m\omega}{\hbar}\rho_<^2\right), \quad (246)$$

where $k_b = E_b/(2\hbar\omega)$. Since $G(\rho,\rho'; l_b, E_b)$ is not a multi-valued function of E_b, it has no branch point on the E_b-plane and contains no contribution corresponding to a continuous spectrum, but has poles at $k_b = \nu + \frac{1}{2}L_b + \frac{1}{2}$ with $\nu \in \mathbb{N}_0$ yielding the discrete energy spectrum (225).

Finally, we compare the Green function for the bound state of the Coulomb problem (244) and the Green function for the radial oscillator (246). The Gamma functions and the Whittaker functions in (244) are brought to those in (246) by transformations $r = C\rho^2$ with $C = m\omega/(2\hbar\kappa)$, $L_a = L_b/\eta$ with $\eta = 2$, and $k_a = k_b$. Although the first two transformations are two of the dual operations, the last one must be verified. Since $k_a = me^2/(\hbar^2\kappa) = -m\lambda_a/(\hbar^2\kappa)$ and $\lambda_a - -E_b/(4C)$, it immediately follows that $k_a - E_b/(2\hbar\omega) = k_b$ provided $C = m\omega/(2\hbar\kappa)$. For the bound state problem, $k_a = \nu + L_a + \frac{1}{2}$ and $k_b = \nu + \frac{1}{2}L_b + \frac{1}{2}$. Hence, it is apparent that $k_a = k_b$ when $L_a = L_b/2$. The extra function in (219) is now given by $h(\rho)h(\rho') = \sqrt{m\omega/(\hbar\kappa)}\sqrt{\rho\rho'}$. Hence the prefactor $m/(\hbar^2\kappa)$ in (244) divided by the extra function gives rise to the prefactor $(\hbar\omega\sqrt{\rho\rho'})^{-1}$ in (246). In this fashion, $G_a(r,r'; L_a, E_a)$ of (244) is completely transformed into $G_b(\rho,\rho'; L_b, E_b)$ by the duality procedures with $C = (m\omega/2\hbar\kappa)$. By letting $L_b = \ell_b + D/2 - 1$ with $\ell_b \in \mathbb{N}_0$, we can see that the formula (221) works well for the Coulomb–Hooke pair.

4.4. A Confinement Potential as a Multi-Term Power-Law Example

One of the motivations that urged the study of power-law potentials was the quark-antiquark confinement problem. See, for instance, references [36,39,40]. Here we examine a two-term power potential as a model of the confinement potential.

Let system A consist of a particle of mass m confined in a two-term power potential,

$$V_a(r) = \lambda_a r^a + \lambda_{a'} r^{a'}, \tag{247}$$

where $\lambda_a \neq 0$, $\lambda_{a'} \neq 0$, $a \neq a'$, $a \neq 0$, and $a' \neq 0$. Let system B be power-dual to system A and quantum-mechanically solvable. Then we expect that some quantum-mechanical information can be obtained concerning the confined system A by analyzing the properties of system B. As we have seen earlier, when system A and system B are dual to each other, the shifted potential of system A,

$$U_a(r) = \lambda_a r^a + \lambda_{a'} r^{a'} - E_a, \tag{248}$$

transforms to that of system B,

$$U_b(\rho) = \lambda_b \rho^b + \lambda_{b'} \rho^{b'} - E_b, \tag{249}$$

by

$$U_b(\rho) = g(\rho) U_a(f(\rho)). \tag{250}$$

Here $r = f(\rho) = C\rho^\eta$, $g(\rho) = C^2 \eta^2 \rho^{2\eta - 2}$, $\eta = 2/(a+2) = -b/a$, and

$$b' = 2(a' - a)/(a + 2) \qquad \lambda_{b'} = \lambda_{a'} \eta^2 C^{a'+2}. \tag{251}$$

Note also that the exchange relations,

$$E_b = -\eta^2 C^{a+2} \lambda_a, \qquad \lambda_b = -\eta^2 C^2 E_a, \tag{252}$$

play an essential role in verifying the equality (250).

First, we wish to tailor the potential of system A to be a confinement potential. To this end, we set the following conditions.

(i) System B behaves as a radial harmonic oscillator $(\lambda_b = 0, \lambda_{b'} > 0, b' = 2)$

(ii) System A has a bound state with $E_a = 0$ and its potential is asymptotically linearly-increasing $(\lambda_{a'} > 0, a' = 1)$.

Since we are unable to solve analytically the Schrödinger equation for system B with (249) in general, we consider the limiting case for which $\lambda_b \to 0$, that is, we employ for the potential of system B

$$\mathcal{U}_b(\rho) = \lim_{\lambda_b \to 0} U_b(\rho) = \lambda_{b'} \rho^{b'} - E_b. \tag{253}$$

According to the second relation of (252), the limit $\lambda_b \to 0$ implies $E_a \to 0$. Hence we study only the zero-energy state of system A by assuming that it exists and is characterized by an integral number ν_0. We denote the zero-energy by $E_a(\nu_0)$. There are only a few exactly soluble nontrivial examples with $\mathcal{U}_b$ of (253). Our choice is the one for the radial harmonic oscillator with $b' = 2$ and $\lambda_{b'} > 0$,

$$\mathcal{U}_b(\rho) = \lambda_{b'} \rho^2 - E_b \quad (\lambda_{b'} > 0). \tag{254}$$

Namely, we consider that system B behaves as the radial harmonic oscillator with frequency $\Omega = \sqrt{2\lambda_{b'}/m}$ and angular momentum L_b. Since $b' = 2$ implies $2(a' - a)/(a + 2) = 2$ as obvious from (251), the corresponding potential of system A is

$$V_a(r) = \lambda_a r^{(a'-2)/2} + \lambda_{a'} r^{a'}. \tag{255}$$

Next we assume that a possible confinement potential behaves asymptotically as a linearly increasing function. Thus, letting $a' = 1$ and $\lambda_{a'} > 0$ in (255), we have

$$V_a(r) = \lambda_a r^{-1/2} + \lambda_{a'} r, \qquad (\lambda_a < 0, \, \lambda_{a'} > 0). \tag{256}$$

If $\lambda_a > 0$, then $V_a(r) > 0$ for all r, and the assumed zero-energy state cannot exist. For $\lambda_a < 0$, the effective potential of system A,

$$V_a^{eff}(r) = \frac{(L_a^2 - \frac{1}{4})\hbar^2}{2mr^2} - |\lambda_a| r^{-1/2} + |\lambda_{a'}| r, \tag{257}$$

can accommodate the zero-energy state provided that λ_a and $\lambda_{a'}$ are so selected that $V_a^{eff}(r_1) < 0$ where r_1 is a positive root of $dV_a^{eff}(r)/dr = 0$. Here $L_a = L_b/\eta$ and $L_a = \ell_a + D/2 - 1$ with $\ell_a \in \mathbb{N}_0$. In this manner, we are able to obtain the confinement potential (256) which is asymptotically linearly increasing and may accommodate at least the assumed zero-energy state. Figure 2 shows the effective potential (257) of system A for $\ell_a = 1, D = 3, \lambda_{a'} = 1$ and $v_0 = 0, 1, 2, 3, 4$ in units $2m = \hbar = 1$.

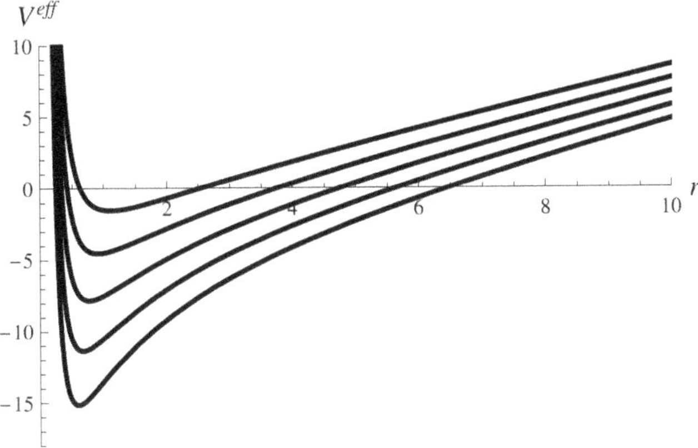

Figure 2. The effective potential (257) related to the eigenfunctions (266) for $v_0 = 0, 1, 2, 3, 4$ from top to bottom. The parameters and units are set to $\ell_a = \lambda_{a'} = 1, D = 3$ and $2m = \hbar = 1$, respectivly.

Since $a' = 1$, we have $a = (a' - 2)/2 = -1/2$, $\eta = 2/(a + 2) = 4/3$ and $b = -a\eta = 2/3$. The last information concerning b is unimportant insofar as $\lambda_b \to 0$ is assumed. The second relation of (251) demands that

$$C = (9\lambda_{b'}/16\lambda_{a'})^{1/3}. \tag{258}$$

Therefore, the first relation of (252) yields

$$E_b = -\frac{4}{3}\lambda_a \sqrt{\frac{\lambda_{b'}}{\lambda_{a'}}}. \tag{259}$$

On the other hand, since system B behaves as a radial harmonic oscillator with frequency $\Omega = \sqrt{2\lambda_{b'}/m}$ and angular momentum L_b, its energy spectrum is given by

$$E_b(v_0, \ell_b) = \hbar\Omega\,(2v_0 + L_b + 1), \tag{260}$$

where $v = v_0$ is fixed by $E_a(v_0)$ and $L_b = \ell_b + D/2 - 1$ with $\ell_b \in \mathbb{N}_0$. Letting $L_b = (4/3)L_a$ in (260) and interpreting that E_b of (259) represents an allowed value in the spectrum (260), we observe that the coupling constant λ_a may take one of the values specified by the set of (v_0, ℓ_a) via

$$\lambda_a = -\frac{3}{4} \sqrt[4]{\frac{2\lambda_{a'}\hbar^2}{m}} \left(2v_0 + (4/3)L_a + 1\right), \tag{261}$$

where $L_a = \ell_a + D/2 - 1$ with $\ell_a \in \mathbb{N}_0$.

The energy eigenfunction of the radial oscillator has been given in (234). Replacing $(m\omega/\hbar)$ in the previous result by $\beta = m\Omega/\hbar = \sqrt{2m\lambda_{b'}}/\hbar$, we write down the eigenfunction of the present oscillator as

$$\phi_b(\rho; L_b, v_0) = \mathcal{N}_b(L_b, v_0, \beta) \left(\beta\rho^2\right)^{-1/4} M_{v_0 + \frac{1}{2}L_b + \frac{1}{2}, \frac{1}{2}L_b}\left(\beta\rho^2\right), \tag{262}$$

which is normalized to unity with

$$\mathcal{N}_b(L_b, v_0, \beta) = \frac{(4\beta)^{1/4}}{\Gamma(L_b + 1)} \sqrt{\frac{\Gamma(v_0 + L_b + 1)}{\Gamma(v_0 + 1)}}. \tag{263}$$

Moreover, utilizing the eigenfunction just obtained, we construct the eigenfunction for the zero-energy state in the confinement potential (256) by following the simple prescription $\phi_a(r) = h(\rho)\phi_b(\rho)$. For the pair $(a, b) = (-1/2, 2/3)$, the two variables r and ρ are related by $r = C\rho^{4/3}$ with C given in (258). Since $\rho^2 = C^{-3/2}r^{3/2}$ and $C^{-3/2} = (4/3)\sqrt{\lambda_{a'}/\lambda_{b'}}$, we let

$$\alpha = \frac{4}{3}\beta\sqrt{\frac{\lambda_{a'}}{\lambda_{b'}}} = \frac{4}{3}\frac{\sqrt{2m\lambda_{a'}}}{\hbar}, \qquad \beta = \frac{\sqrt{2m\lambda_{b'}}}{\hbar}, \tag{264}$$

and

$$\beta\rho^2 = \alpha r^{3/2}. \tag{265}$$

Multiplying $\phi_b(\rho)$ of (262) by $h(\rho) = \sqrt{dr/d\rho} = \sqrt{4C/3}\rho^{1/6}$, and substituting (264) and $L_b = (4/3)L_a$ into $\phi_b(\rho)$, we arrive at the eigenfunction for the zero-energy state of system A,

$$\phi_a(\rho; L_a, v_0) = \mathcal{N}_a(L_a, v_0, \alpha) \left(\alpha r^{3/2}\right)^{-1/6} M_{v_0 + \frac{2}{3}L_a + \frac{1}{2}, \frac{2}{3}L_a}\left(\alpha r^{3/2}\right), \tag{266}$$

where $L_a = \ell_a + D/2 - 1$ with $\ell_a \in \mathbb{N}_0$. Here the factor $\mathcal{N}_a(L_a, v_0, \alpha)$ that normalizes $\phi_a(\rho)$ to unity cannot be determined by $\mathcal{N}_b((4/3)L_a, v_0, (3/4)\alpha\sqrt{\lambda_{b'}/\lambda_{a'}})$. Corresponding to the value of λ_a specified in (261) by the set (v_0, ℓ_a), the eigenfunction $\phi_a(\rho; \ell_a, v_0)$ is characterized by the same set (v_0, ℓ_a) of numbers.

The Green function of system A obeys the inhomogeneous radial equation,

$$\left\{\frac{d^2}{dr^2} - \frac{L_a^2 - 1/4}{r^2} - \frac{2me^2}{\hbar^2}\left(\lambda_a r^{-1/2} + \lambda_{a'}r\right) + \frac{2me^2}{\hbar^2}E_a\right\} G_a(r, r'; E_a, L_a) = -\frac{2m}{\hbar^2}\delta(r - r'). \tag{267}$$

Since the Green function for the radial oscillator has been given in (246), we can write down the Green function $G_b(\rho, \rho'; E_b(v_0))$ of system B with $\lambda_b = 0$ as

$$G_b(\rho, \rho'; E_b, L_b) = -\frac{m}{\hbar^2\beta} \frac{1}{\sqrt{\rho\rho'}} \frac{\Gamma(\frac{1}{2}L_b - k_b + \frac{1}{2})}{\Gamma(L_b + 1)} W_{k_b, \frac{1}{2}L_b}(\beta\rho_>^2) M_{k_b, \frac{1}{2}L_b}(\beta\rho_<^2), \tag{268}$$

where $k_b = E_b/(2\hbar\Omega)$. The pole of $G_b(\rho, \rho'; E_b)$ that corresponds to $E_b(v_0)$ occurs when $k_b(L_b, v_0) = v_0 + \frac{1}{2}L_b + \frac{1}{2}$ where v_0 is a non-negative integer.

The Green function $G_a(r, r'; E_a, L_a)$ of system A at $E_a = 0$ can be found by substituting (265) together with

$$h(\rho) = \sqrt{4/3}C^{3/8}r^{1/8}, \qquad \frac{1}{2}L_b = \frac{2}{3}L_a,$$

into $h(\rho)h(\rho')G_b(\rho,\rho';E_b,L_b)$. Namely,

$$G_a(r,r';E_a=0,L_a) = \frac{4}{3}C^{3/4}(rr')^{1/8}\,G_b\!\left((r/C)^{3/4},(r'/C)^{3/4};E_b=\frac{16}{9}|\lambda_a|,\frac{4}{3}L_a\right),\quad(269)$$

where C has been given in (258). Explicitly, we have

$$G_a(r,r';E_a,L_a) = -\frac{4m}{3\hbar^2\alpha}(rr')^{-1/4}\frac{\Gamma(\frac{2}{3}L_a-k_a+\frac{1}{2})}{\Gamma(\frac{4}{3}L_a+1)}\,W_{k_a,\frac{2}{3}L_a}(\alpha r_>^{3/2})\,M_{k_a,\frac{3}{2}L_a}(\alpha r_<^{3/2}).\quad(270)$$

where α and β have been given by (264). The pole corresponding to $E_a(\nu_0)=0$ occurs when $k_a = \nu_0 + (2/3)L_a + 1/2$ and $L_a = \ell_a + D/2 - 1$. We have to remember that the Green function (269) is meaningful only in the vicinity of $E_a = 0$.

Remark 25. *The angular momentum L in (167) is identical in form to that used in the semiclassical case (86). However, no Langer-like ad hoc treatment has been made in the Schrödinger equation. The angular contribution $\ell(\ell+D-2)$ and an additional contribution $(D-1)(D-3)/4$ from the kinetic term due to the transformation of base function, $R_\ell(r)$ to $\psi_\ell(r)$, make up the term $L^2-1/4$ in the effective centrifugal potential term of (169).*

Remark 26. *The time transformation $\mathfrak{T}$ needed in classical mechanics takes no part in the power duality of the stationary Schrödinger equation. Instead, the change of the base function plays an essential role. While $\mathfrak{T}$ assumes $dt = g(\rho)ds$, the state function changes as $\psi_a(r) = [g(\rho)]^{1/4}\psi_b(\rho)$. The possible connection between the time transformation and the change of state function has been discussed in the context of path integration for the Green function in [41]. So long as the stationary Schrödinger equation is concerned, there is no clue to draw any causal relation between $\mathfrak{T}$ and $\mathfrak{F}$. However, one might expect that $\mathfrak{T}$ would play a role in the time-dependent Schödinger equation. If the energy-coupling exchange operation $\mathfrak{E}$ of (93) is formally modified as*

$$\mathfrak{E}' : gV_a(r) \to -i\hbar\frac{\partial}{\partial \bar{s}},\qquad gi\hbar\frac{\partial}{\partial \bar{t}} \to -V_b(\rho),\quad(271)$$

then the time-dependent radial Schrödinger equation,

$$\left[-\frac{\hbar^2}{2m}\frac{d^2}{dr^2} + \frac{\hbar^2(L_a^2-1/4)}{2mr^2} + V_a(r)\right]\psi_a(r) = i\hbar\frac{\partial\psi_a(r)}{\partial\bar{t}},\quad(272)$$

transforms into

$$\left[-\frac{\hbar^2}{2m}\frac{d^2}{d\rho^2} + \frac{\hbar^2(L_b^2-1/4)}{2m\rho^2} + V_b(\rho)\right]\psi_b(\rho) = i\hbar\frac{\partial\psi_b(\rho)}{\partial\bar{s}},\quad(273)$$

under the set of $\{\mathfrak{R}, \mathfrak{L}, \mathfrak{E}', \mathfrak{F}\}$. It is important that $\bar{t}$ and $\bar{s}$ are not necessarily connected by $\mathfrak{T}$; they are basically independent time-like parameters. In conclusion, the time transformation $\mathfrak{T}$ has no role in the time-dependent Schrödinger equation.

Remark 27. *More on time transformations. Since we are dealing with the action integral (170) rather than the Schrödinger equation, it is easy to observe that the time transformation $\mathfrak{T}$ in the classical action in Section 2 is closely related to the transformation $\mathfrak{F}$ of wave functions in the quantum action (170). Recall that $\mathfrak{T} : (dt/d\varphi) = g(\rho)(ds/d\varphi)$ where $g = f'^2$ with $f = C\rho^\eta$, and that*

$$dt\,U_a = ds\,gU_a = ds\,U_b.\quad(274)$$

From (171) and (190), we have

$$dr\,U_a\psi_a^*\psi_a = d\rho\,f'h^2U_a\psi_b^*\psi_b = d\rho\,U_b\psi_b^*\psi_b,\quad(275)$$

where $g = f'h^2 = f'^2$. Comparing (274) and (275), we see that $dt = g\,ds$ in classical mechanics corresponds to $dr\,\psi_a^*\psi_a = g\,d\rho\,\psi_b^*\psi_b$ in quantum mechanics. In other words, $dr\,\psi_a^*\psi_a$ has the same transformation behavior that dt does. In this respect, we may say that the role of $\mathfrak{T}$ in classical mechanics is replaced by $\mathfrak{F}$ in quantum mechanics.

5. Summary and Outlook

In the present paper we have revisited the Newton–Hooke power-law duality and its generalizations from the symmetry point of view.

(1) We have stipulated the power-dual symmetry in classical mechanics by form-invariance and reciprocity of the classical action in the form of Hamilton's characteristic function, and clarified the roles of duality operations $\{\mathfrak{C}, \mathfrak{R}, \mathfrak{T}, \mathfrak{E}, \mathfrak{L}\}$. The exchange operation $\mathfrak{E}$ has a double role; it may decide the constant C appearing in the transformation $r = C\rho^{\eta}$, while it leads to an energy formula that relates the new energy to the old energy.

(2) We have shown that the semiclassical action is symmetric under the set of duality operations $\{\mathfrak{C}, \mathfrak{R}, \mathfrak{E}, \mathfrak{L}\}$ without $\mathfrak{T}$ insofar as angular momentum L is treated as a continuous parameter, and observed that the power-duality is essentially a classical notion and breaks down at the level of angular quantization. To preserve the basic spirit of power-duality in the semiclassical action, we have proposed an ad hoc procedure in which angular momentum transforms as $L_b = \eta L_a$, as the classical case, rather than $\ell_b = \eta \ell_a$; after that each of L is quantized as $L = \ell + D/2 - 1$ with $\ell \in \mathbb{N}_0$. As an example, we have solved by the WKB formula a simple problem for a linear motion in a fractional power potential.

(3) We have failed to verify the dual symmetry of the supersymmetric (SUSY) semiclassical action for an arbitrary power potential, but have succeeded to reveal the Coulomb–Hooke duality in the SUSY action.

(4) To study the power-dual symmetry in quantum mechanics, we have chosen the action in which the variables are the wave function $\psi(r)$ and its complex conjugate $\psi^*(r)$ and from which the radial Schrödinger equation can be derived. The potential appearing in the action is a two-term power potential. We have shown that the action is symmetric under the set of operations $\{\mathfrak{C}, \mathfrak{R}, \mathfrak{E}, \mathfrak{L}\}$ plus the transformation of wave function $\mathfrak{F}$ provided that angular momentum L is a continuous parameter. Again the ad hoc procedure introduced for the semiclassical case must be used in quantum mechanics. Associated with $\mathfrak{F}$ is the transformation of Green functions from which we have derived a formula that relates the new Green function and the old one. We have studied the Coulomb–Hooke duality to verify the energy formula and the formula for the Green functions. We also discussed a confinement potential and the Coulomb–Hooke–Morse triality.

There are more topics that we considered important but left out for the future work. They include the power-dual symmetry in the path integral formulation of quantum mechanics, the Coulomb–Hooke duality in Dirac's equation, and the confinement problem in Witten's framework of supersymetric quantum mechanics. Feynman's path integral is defined for the propagator (or the transition probability) with the classical action in the form of Hamilton's principal function, whereas the path integral pertinent to the duality discussion is based on the classical action in the form of Hamilton's characteristic function. Since the power-dual symmetry of the characteristic action has been shown, it seems obvious that the path integral remains form-invariant under the duality operations, but the verification of it is tedious. As is well-known, Dirac's equation is exactly solvable for the hydrogen atom. There are also solutions of Dirac's equation for the harmonic oscillator. However, the Coulomb–Hooke duality of Dirac's equation has never been established. The situation is similar to Witten's model of SUSYQM. Using the same superpotential as that used for the semiclassical case in Section 4, we may be able to show the Coulomb–Hooke symmetry and handle the confinement problem in Witten's framework.

Author Contributions: All authors contributed equally to the design and methodology of this research, to the analysis of the results and to the writing and reviewing of the manuscript. All authors have read and agreed to the published version of the manuscript.

Funding: This research received no external funding.

Conflicts of Interest: The authors declare no conflict of interest.

Appendix A. The Coulomb–Hooke–Morse Triality

In this Appendix A, we wish to present the Coulomb–Hooke–Morse triality that relates the Morse oscillator to the Coulomb–Hooke duality. Specifically, letting system A be the hydrogen atom (for the Coulomb system), system B be the radial harmonic oscillator (for the Hooke system) and system C be the Morse oscillator, we deal with their triangular relation. The Morse oscillator is a system obeying the one-dimensional Schrödinger Equation [71],

$$-\frac{\hbar^2}{2m}\frac{d^2\psi_c(\xi)}{d^2\xi} + (V_c(\xi) - E_c)\psi_c(\xi) = 0, \quad \xi \in \mathbb{R}, \tag{A1}$$

where

$$V_c(\xi) = D_1\,e^{-2\alpha\xi} - 2D_2\,e^{-\alpha\xi}, \qquad \alpha,\,D_1,\,D_2 > 0, \tag{A2}$$

which is the Morse potential in a slightly modified form. The potential (A2), being not a power-law potential, is beyond the scope of the main text. It is yet interesting to observe how the Morse oscillator is related to the Coulomb–Hooke duality. It is straightforward, if one follows the general transformation procedure [41] for the Schrödinger equation, to transform (A1) directly to the Schrödinger equation for each of the hydrogen atom and the radial harmonic oscillation. Here, to focus our attention on their trial nature, we place the Whittaker function at the center of the triangular relation. In fact, the Schrödinger Equation (A1) is easily transformed to the Whittaker Equation (226) under the substitutions

$$x = \gamma\,e^{-\alpha\xi}, \quad \gamma = \frac{\sqrt{8mD_1}}{\hbar\alpha} \tag{A3}$$

$$L_c = \frac{\sqrt{-2mE_c}}{\hbar\alpha}, \quad k_c = \sqrt{\frac{2mD_2^2}{\hbar^2\alpha^2 D_1}}, \tag{A4}$$

$$w(x) = x^{1/2}\psi_c(\xi). \tag{A5}$$

Hence the bound state solution of (A1) can be expressed in terms of the Whittaker function as

$$\psi_c(\xi) = \mathcal{N}_c\,e^{\alpha\xi/2}M_{k_c,L_c}\!\left(\gamma\,e^{-\alpha\xi}\right), \tag{A6}$$

subject to the condition

$$k_c = \nu + L_c + \frac{1}{2}, \quad \nu \in \mathbb{N}_0. \tag{A7}$$

The last condition yields the energy spectrum,

$$E_c = -\frac{\hbar^2\alpha^2}{2m}\left\{\sqrt{\frac{2mD_2^2}{\hbar^2\alpha^2 D_1}} - \left(\nu + \frac{1}{2}\right)\right\}^2, \quad \nu = 0,1,2,\ldots < \sqrt{\frac{2mD_2^2}{\hbar^2\alpha^2 D_1}} - \frac{1}{2}. \tag{A8}$$

The Morse oscillator solution $\psi_c(\xi)$ in (A6) may be compared with the Coulomb bound state solution $\psi_a(r)$ and the Hooke oscillator solution $\psi_b(\rho)$ given, respectively, by

$$\psi_a(r) = \mathcal{N}_a\,M_{k_a,L_a}(2\kappa r), \tag{A9}$$

with

$$k_a = \nu + L_a + \frac{1}{2}, \quad \nu \in \mathbb{N}_0, \tag{A10}$$

and

$$\psi_b(\rho) = \mathcal{N}_b\left(\frac{m\omega}{\hbar}\rho^2\right)^{-1/4}M_{k_b,\frac{1}{2}L_b}\!\left(\frac{m\omega}{\hbar}\rho^2\right), \tag{A11}$$

with

$$k_b = \nu + \frac{1}{2}L_b + \frac{1}{2} \quad \nu \in \mathbb{N}_0. \tag{A12}$$

The bound state conditions (A10) and (A12) lead to the energy spectrum of the Coulomb system (A) and that of the Hooke system (B), respectively, when

$$k_a = me^2/(\hbar^2\kappa), \quad \hbar\kappa = \sqrt{-2mE_a}, \quad L_a = \ell + 1/2, \quad \ell \in \mathbb{N}_0, \tag{A13}$$

$$k_b = E_b/(\hbar\omega), \quad L_b = \ell + 1/2, \quad \ell \in \mathbb{N}_0. \tag{A14}$$

The triality relations are schematically shown below,

$$
\begin{array}{ccccc}
& \text{Morse} & & & \text{Morse} \\
CA \nearrow & & \nwarrow BC \qquad & AC \nearrow & & \searrow CB \\
\text{Coulomb} & \xrightarrow{AB} & \text{Hooke} \qquad & \text{Coulomb} & \xleftarrow{BA} & \text{Hooke}
\end{array}
$$

and the dual transformations AC, CB and BA are given by

$$
\begin{array}{llll}
AC: & 2\kappa r = \gamma e^{-\alpha\xi}, & k_a = k_c, \quad L_a = L_c, & \psi_a(r) = e^{-\alpha\xi/2}\psi_c(\xi) \\
CB: & \gamma e^{-\alpha\xi} = (m\omega/\hbar)\rho^2, & k_c = k_b, \quad L_c = (1/2)L_b, & e^{-\alpha\xi/2}\psi_c(\xi) = \rho^{-1/2}\psi_b(\rho) \\
BA: & (m\omega/\hbar)\rho^2 = 2\kappa r, & k_b = k_a, \quad (1/2)L_b = L_a, & \rho^{-1/2}\psi_b(\rho) = \psi_a(r)
\end{array}
$$

which are all invertible. Although none of the energy formulas discussed earlier for the power-duality works when the Morse (non-power-law) potential is involved, transforming one of the bound state conditions to another suffices as each condition generates an energy spectrum. Let $\chi(k_s, \eta_s L_s)$ represent the condition $k_s - \eta_s L_s - \frac{1}{2} = \nu$ where $s = a, b, c$, and $\eta_a = \eta_c = 1$ and $\eta_b = 1/2$. The map $\chi(k_s, \eta_s L_s) \Rightarrow \chi(k_{s'}, \eta_{s'} L_{s'})$ induces $E_s \Rightarrow E_{s'}$.

$$
\begin{array}{ccccccccc}
& \chi(k_c, L_c) & & & & & & E_c & \\
CA \nearrow & & \nwarrow BC & & \Rightarrow & & CA \nearrow & & \nwarrow BC \\
\chi(k_a, L_a) & \xrightarrow{AB} & \chi(k_b, \tfrac{1}{2}L_b) & & & & E_a & \xrightarrow{AB} & E_b
\end{array}
$$

Finally, it must be mentioned that this triangular relation has been discussed in the context of so-called shape invariant potentials in supersymmetric quantum mechanics [72]. It may also be worth pointing out that the three systems share the $SU(1,1)$ dynamical group [50,56].

References

1. Mayor, M.; Lovis, C.; Santos, N.C. Doppler spectroscopy as a path to the detection of Earth-like planets. *Nature* **2014**, *513*, 328–335;
2. Pepe, F.; Cristiani, S.; Rebolo, R.; Santos, N.C.; Dekker, H.; Cabral, A.; Di Marcantonio, P.; Figueira, P.; Lo Curta, G.; Lovis, C.; et al. ESPRESSO at VLT, On-sky performance and first results. *Astron. Astrophys.* **2021**, *645*, 26.
3. Lissauer, J.J.; Dawson, R.I.; Tremaine, S. Advances in exoplanet science from Kepler. *Nature* **2014**, *513*, 336–344.
4. Borucki, W.J. Kepler: A Brief Discussion of the Mission and Exoplanet Results. *Proc. Am. Philos. Soc.* **2017**, *161*, 38–65.
5. Lascar, J. On the Spacing of Planetary Systems. *Phys. Rev. Lett.* **2000**, *84*, 3240–3243.
6. Waldvogel, J. Fundamentals of regularization in celestial mechanics and linear perturbation theories. In *Extra-Solar Planets: The Detection, Formation, Evolution and Dynamics of Planetary Systems*; Steves, B.A., Hendry, M., Cameron, A.C., Eds.; Taylor & Francis: Boca Raton, FL, USA, 2010; pp. 169–184.
7. Levi-Civita, T. Sur la résolution qualitative du problème restreint des trois corps. *Acta Math.* **1906**, *30*, 305–327.
8. Levi-Civita, T. Sur la régularisation du problème des trois corps. *Acta Math.* **1920**, *42*, 99–144.
9. Kustaanheimo, P.; Stiefel, E.L. Perturbation theory of Kepler motion based on spinor representation. *J. Reine Angew. Math.* **1965**, *218*, 204–219.
10. Arnold, V.I. *Huygens and Barrow, Newton and Hooke*; Birkhäuser Verlag: Basel, Switzerland, 1990; pp. 95–100.
11. Szebehely, V. *Theory of Orbits, the Restricted Problem of Three Bodies*; Academic Press: New York, NY, USA, 1967.
12. Stiefel E.L.; Scheifele, G. (Eds.) *Linear and Regular Celestial Mechanics*; Springer: Berlin, Germany, 1971.
13. Sundman K.F. Recherches sur le problème des trois corps. *Acta Soc. Sci. Fenn.* **1907**, *34*, 6.

14. Bohlin, M. K. Note sur le problème des deux corps et sur une intégration nouvelle dans le problème des trois corps. *Bull. Astron.* **1911**, *28*, 113.
15. Vivarelli, M.D. The KS transformation in hypercomplex form. *Celest. Mech. Dyn. Astron.* **1983**, *29* 45–50.
16. Vrbik, J. Celestial mechanics via quaternions. *Can. J. Phys.* **1994**, *72*, 141–146.
17. Saari, D.G. A visit to the Newton n-body problem via elementary complex variables. *Am. Math. Monthly* **1990**, *97*, 105–119.
18. Chandrasekhar, S. Chapter 6: Supplement: On dual laws of centripedal attraction. In *Newton's Principia for the Common Reader*; Clarendon Press: Oxford, UK, 1995; pp. 114–125.
19. Janin, G.; Bond, V.R. The Elliptic Anomaly. NASA Technical Memorandum NASA-TM-58228. 1980. p. 20. Available online: https://ntrs.nasa.gov/citations/19800013899 (accessed on 2 March 2021).
20. Nacozy, P. The intermediate anomaly. *Celest. Mech.* **1977**, *16*, 309–313.
21. Ferrándiz, J.M.; Ferrer, S.; Sein-Echaluce, M.L. Generalized elliptic anomalies. *Celest. Mech.* **1987**, *40* 315–328.
22. Needham, T. Newton and the transmutation of force. *Am. Math. Mon.* **1993**, *100*, 119–137.
23. Needham, T. *Visual Complex Analysis*; Clarendon Press: Oxford, UK, 1997; p. 246.
24. Grant, A.K.; Rosner, J.L. Classical orbits in power-law potentials. *Am. J. Phys.* **1994**, *62*, 310–315.
25. Kasner, E. Differential-Geometric Aspects of Dynamics. In *Princeton Colloquium Lectures*; American Mathematical Society: New York, NY, USA, 1913; p. 1934.
26. Hall, R.W.; Josić, K. Planetary motion and the duality of force laws. *SIAM Rev.* **2000**, *42*, 115–124.
27. Grandati, Y.; Berard, A.; Mohrbach, H. Bohlin-Arnold-Vassiliev's duality and conserved quantities. *arXiv* **2008**, arXiv:0803.2610v2.
28. Schrödinger, E. Quantisierung als Eigenwertproblem. *Ann. Phys.* **1926**, *384*, 361–376.
29. Schrödinger,E. Quantisierung als Eigenwertproblem II. *Ann. Phys.* **1926**, *384*, 489–527.
30. Manning, M.F. Exact Solutions of the Schrödinger Equation. *Phys. Rev.* **1935**, *48*, 161–164.
31. Fock, V. Zur Theorie des Wasserstoffatoms. *Z. für Phys.* **1935**, *98*, 145–154.
32. Bargmann, V. Zur Theorie des Wasserstoffatoms. *Z. für Phys.* **1936**, *99*, 576–582.
33. Laporte, O.; Rainich, G.Y. Stereographic parameters and pseudo-minimal hypersurfaces. *Trans. Am. Math. Soc.* **1936**, *39*, 154–182.
34. Jauch, J.M.; Hill, E.L. On the problem of degeneracy in quantum mechanics. *Phys. Rev.* **1940**, *57*, 641–645.
35. Schrödinger, E. Further Studies on Solving Eigenvalue Problems by Factorization. *Proc. Roy. Irish Acad.* **1941**, *46 A*, 183–206.
36. Johnson, B.R. On a connection between radial Schrödinger equations for different power-law potentials. *J. Math. Phys.* **1980**, *21*, 2640–2647.
37. Kostelecký, V.A.; Nieto, M.M.; Truax, D.R. Supersymmetry and the relationship between the Coulomb and oscillator problems in arbitrary dimensions. *Phys. Rev. D* **1985**, *32*, 2627–2633.
38. Kibler, M. Connection between the hydrogen atom and the harmonic oscillator: The zero-energy case. *Phys. Rev. A* **1984**, *29*, 2891–2894.
39. Quigg, C.; Rosner, J.L. Quantum mechanics with applications to quarkonium. *Phys. Rep.* **1979**, *56*, 167–235.
40. Gazeau, J.P. A remarkable duality in one-particle quantum mechanics between some confining potentials and $(R + L_c^\infty)$ potentials. *Phys. Lett. A* **1980**, *75*, 159–163.
41. Junker, G. Remarks on the local time rescaling in path integration. *J. Phys. A* **1990**, *23*, L881–L884.
42. Feynman, R.P.; Hibbs, A.R. *Quantum Mechanics and Path Integrals*; McGraw-Hill: New York, NY, USA, 1965.
43. Schulman, L.S. *Techniques and Applications of Path Integration*; Dover Publ.: New York, NY, USA, 2005.
44. Duru, I.H.; Kleinert, H. Solution of the path integral for the H-atom. *Phys. Lett. B* **1979**, *84*, 185–188.
45. Ho, R.; Inomata, A. Exact path integral treatment of the hydrogen atom. *Phys. Rev. Lett.* **1982**, *48*, 231–234.
46. Peak, D.; Inomata, A. Summation over Feynman histories in polar coordinates. *J. Math. Phys.* **1969**, *10*, 1422–1428.
47. Inomata, A. Alternative exact-path-integral-treatment of the hydrogen atom. *Phys. Lett. A* **1984**, *101*, 253–257.
48. Steiner, F. Space-time transformations in radial path integrals. *Phys. Lett. A* **1984**, *106*, 356–362.
49. Steiner, F. Exact path integral treatment of the hydrogen atom. *Phys. Lett. A* **1984**, *106*, 363–367.
50. Inomata, A.; Kuratsuji, H.; Gerry, C.C. *Path Integrals and Coherent States of SU(2) and SU(1, 1)*; World Scientific: Singapore, 1992.
51. Grosche, C.; Steiner, F. *Handbook of Feynman Path Integrals*; Springer: Berlin, Germany, 1998.
52. Barut, A.O.; Inomata, A.; Junker, G. Path integral treatment of the hydrogen atom in a curved space of constant curvature. *J. Phys. A* **1987**, *20*, 6271–6280.
53. Barut, A.O.; Inomata, A.; Junker, G. Path integral treatment of the hydrogen atom in a curved space of constant curvature: II. Hyperbolic space. *J. Phys. A* **1990**, *23*, 1179–1190.
54. Inomata, A.; Junker, G. Path integral quantization of Kaluza-Klein monopole systems. *Phys. Rev. D* **1991**, *43*, 1235–1242.
55. Böhm, M.; Junker, G. Path Integration over Compact and Noncompact Rotation Groups. *J. Math. Phys.* **1987**, *28*, 1978–1994.
56. Inomata, A.; Junker, G. Path integrals and Lie groups. In *Noncompact Lie Groups and Some of Their Applications*; Tanner, E.A., Wilson, R., Eds.; Kluwer: Dordrecht, The Netherlands, 1994; pp. 199–224.
57. Steiner, F. Path integrals in polar co-ordinates from eV to GeV. In *Path Integrals from meV to MeV*; Gutzwiller, M.C., Inomata, A., Klauder, J.R., Streit, L., Eds.; World Scientific: Singapore, 1986; pp. 335–359.
58. Atiyah, M. *Duality in Mathematics and Physics*; Lecture Notes; Institut de Matemàtica de la Universitat de Barcelona: Barcelona, Spain, 2007.

59. Comtet, A.; Bandrauk, A.D.; Campbell, D.K. Exactness of semiclassical bound state energies for supersymmetric quantum mechanics. *Phys. Lett. B* **1985**, *150*, 159–162.
60. Eckhardt, B. Maslov-WKB theory for supersymmetric hamiltonians. *Phys. Lett. B* **1986**, *168*, 245–247.
61. Inomata, A.; Junker, G. Quasiclassical path-integral approach to supersymmetric quantum mechanics. *Phys. Rev. A* **1994**, *50*, 3638–3649.
62. Junker, G. *Supersymmetric Methods in Quantum, Statistical and Solid State Physics, Enlarged and Revised Edition*; IOP Publishing: Bristol, UK, 2019.
63. Louck, J.D. Generalized orbital angular momentum and the n-fold degenerate quantum-mechanical oscillator: Part III. Radial integrals. *J. Mol. Spectrosc.* **1960**, *4*, 334–341.
64. Bergmann, D.; Frishman, Y. A Relation between the Hydrogen Atom and Multidimensional Harmonic Oscillators. *J. Math. Phys.* **1965**, *6*, 1855.
65. Kostecký, V.A.; Russell, N. Radial Coulomb and oscillator systems in arbitrary dimensions. *J. Math. Phys.* **1996**, *37*, 2166–2181.
66. Louck, J.D.; Shaffer, W.H. Generalized orbital angular momentum and the n-fold degenerate quantum-mechanical oscillator: Part I. The twofold degenerate oscillator. *J. Mol. Spectrosc.* **1960**, *4*, 285–297.
67. Langer, R.E. On the Connection Formulas and the Solutions of the Wave Equation. *Phys. Rev.* **1937**, *51*, 669–676.
68. Ishkhanyan, A.M. Exact solution of the Schrödinger equation for the inverse square root potential $V_0/\sqrt{x}$. *Eur. Phys. Lett.* **2015**, *112*, 10006.
69. Whittaker, E.T.; Watson,G.N. *A Course of Modern Analysis*; Cambridge University Press: Cambridge, UK, 1955; p. 337.
70. Morse, P.M.; Feshbach, H. *Methods of Theoretical Physics*; McGraw-Hill: New York, NY, USA, 1953; p. 926.
71. Morse, P.M. Diatomic molecules according to the wave mechanics. II. Vibrational levels. *Phys. Rev.* **1920**, *34*, 57–64.
72. Cooper, F.; Khare, A.; Sukhatme,U. *Supersymmetry in Quantum Mechanics*; World Scientific: Singapore, 2001; Appendix B.

 *symmetry*

Article

Time-Dependent Conformal Transformations and the Propagator for Quadratic Systems †

Qiliang Zhao [1], Pengming Zhang [1] and Peter A. Horvathy [2,*]

1 School of Physics and Astronomy, Sun Yat-sen University, Zhuhai 519082, China; zhaoqliang@mail2.sysu.edu.cn (Q.Z.); zhangpm5@mail.sysu.edu.cn (P.Z.)
2 Institut Denis Poisson, Tours University-Orléans University, UMR 7013, F-37200 Tours, France
* Correspondence: horvathy@lmpt.univ-tours.fr
† This paper celebrates the 90th birthday of Akira Inomata.

Abstract: The method proposed by Inomata and his collaborators allows us to transform a damped Caldirola–Kanai oscillator with a time-dependent frequency to one with a constant frequency and no friction by redefining the time variable, obtained by solving an Ermakov–Milne–Pinney equation. Their mapping "Eisenhart–Duval" lifts as a conformal transformation between two appropriate Bargmann spaces. The quantum propagator is calculated also by bringing the quadratic system to free form by another time-dependent Bargmann-conformal transformation, which generalizes the one introduced before by Niederer and is related to the mapping proposed by Arnold. Our approach allows us to extend the Maslov phase correction to an arbitrary time-dependent frequency. The method is illustrated by the Mathieu profile.

Keywords: quantum mechanics; semiclassical theories and applications; classical general relativity

PACS: 03.65.-w quantum mechanics; 03.65.Sq semiclassical theories and applications; 04.20.-q classical general relativity

Citation: Zhao, Q.; Zhang, P.; Horvathy, P.A. Time-Dependent Conformal Transformations and the Propagator for Quadratic Systems. *Symmetry* **2021**, *13*, 1866. https://doi.org/10.3390/sym13101866

Academic Editor: Georg Junker

Received: 7 July 2021
Accepted: 17 August 2021
Published: 3 October 2021

Publisher's Note: MDPI stays neutral with regard to jurisdictional claims in published maps and institutional affiliations.

1. Introduction

A nonrelativistic quantum particle with unit mass in $d + 1$ spacetime dimensions with coordinates x, t is given by the natural Lagrangian $L = \frac{1}{2}\dot{x}^2 - V(x, t)$. The wave function is expressed in terms of the propagator,

$$\psi(x'', t'') = \int K(x'', t''|x', t')\psi(x', t')dx' \tag{1}$$

which, following Feynman's intuitive proposal [1], is obtained as,

$$K(x'', t''|x', t') = \int \exp\left[\frac{i}{\hbar}\mathcal{A}(\gamma)\right]\mathcal{D}\bigcirc, \tag{2}$$

where the (symbolic) integration is over all paths $\gamma(t) = (x(t), t)$ that link the spacetime point (x', t') to (x'', t'') and where:

$$\mathcal{A}(\gamma) = \int_{t'}^{t''} L\big(\gamma(t), \dot{\gamma}(t), t\big)dt \tag{3}$$

is the classical action calculated along $\gamma(t)$ [1–3].

The rigorous definition and calculation of (2) are beyond our scope here. However, the *semiclassical approximation* leads to the van Vleck–Pauli formula [2–5],

$$K(x'', t''|x', t') = \left[\frac{i}{2\pi\hbar}\frac{\partial^2 \bar{\mathcal{A}}}{\partial x' \partial x''}\right]^{1/2} \exp\left[\frac{i}{\hbar}\bar{\mathcal{A}}(x'', t''|x', t')\right], \tag{4}$$

where $\bar{A}(x'',t''|x',t') = \int_{t'}^{t''} L(\bar{\gamma}(t),\dot{\bar{\gamma}}(t),t)dt$ is the classical action calculated along the (supposedly unique (This condition is satisfied away from caustics [2,3,6]. Moreover, (5) and (8) are valid only for $0 < T'' - T'$ and for $0 < t'' - t' < \pi$, respectively, as discussed in Section 4.)) classical path $\bar{\gamma}(\tau)$ from (x',t') and (x'',t''). This expression involves data of the classical motion only. We note here also the van Vleck determinant $\frac{\partial^2 \bar{A}}{\partial x' \partial x''}$ in the prefactor [4,5].

Equation (4) is exact for a quadratic-in-the-position potentials in $1+1$ dimension $V(x,t) = \frac{1}{2}\omega^2(t)\,x^2$ that we consider henceforth.

For $\omega \equiv 0$, i.e., for a free nonrelativistic particle of unit mass in $1+1$ dimensions with coordinates X and T, the result is [1–3],

$$K_{free}(X'',T''|X',T') = \left[\frac{1}{2\pi i\hbar(T''-T')}\right]^{1/2} \exp\left\{\frac{i}{\hbar}\frac{(X''-X')^2}{2(T''-T')}\right\}. \tag{5}$$

A harmonic oscillator with dissipation is in turn described by the Caldirola–Kanai (CK) Lagrangian and the equation of motion, respectively [7,8]. For constant damping and a harmonic frequency, we have,

$$L_{CK} = \frac{1}{2}e^{\lambda_0 t}\left(\left(\frac{dx}{dt}\right)^2 - \omega_0^2 x^2\right), \tag{6}$$

$$\frac{d^2 x}{dt^2} + \lambda_0 \frac{dx}{dt} + \omega_0^2 x = 0 \tag{7}$$

with $\lambda_0 = \text{const.} > 0$ and $\omega_0 = \text{const.}$. A lengthy calculation then yields the exact propagator [2,3,9–11]:

$$K_{CK}(x'',t''|x',t') = \left[\frac{\Omega_0\, e^{\frac{\lambda_0}{2}(t''+t')}}{2\pi i\hbar\, \sin\left[\Omega_0(t''-t')\right]}\right]^{\frac{1}{2}} \times \tag{8}$$

$$\exp\left\{\frac{i\Omega_0}{2\hbar\,\sin\left[\Omega_0(t''-t')\right]}\left[(x''^2 e^{\lambda_0 t''} + x'^2 e^{\lambda_0 t'})\cos\left[\Omega_0(t''-t')\right] - 2x''x' e^{\lambda_0 \frac{t''+t'}{2}}\right]\right\},$$

$$\Omega_0^2 = \omega_0^2 - \tfrac{1}{4}\lambda_0^2, \tag{9}$$

where an irrelevant phase factor was dropped.

Inomata and his collaborators [12–15] generalized (9) to a time-dependent frequency by redefining time, $t \to \tau$, which allowed them to transform the time-dependent problem to one with a constant frequency (see Section 2). Then, they followed by what they called a *"time-dependent conformal transformation"* $(x,t) \to (X,T)$ such that:

$$x = f(T)\,X(T)\exp\left[\tfrac{1}{2}\lambda_0 T\right], \quad t = g(T), \quad \text{where} \quad f^2(T) = \frac{dg}{dT}, \tag{10}$$

which allowed them to derive the propagator from the free expression (5). When spelled out, (10) boils down to a generalized version, (22), of the correspondence found by Niederer [16].

It is legitimate to wonder: *in what sense are these transformations "conformal" ?* In Section 3, we explain that, in fact, *both* mappings can be interpreted in the Eisenhart–Duval (E-D) framework as conformal transformations between two appropriate Bargmann spaces [17–21]. Moreover, the change of variables $x,t \to X,T$ is a special case of the one put forward by Arnold [22,23] and is shown to be convenient to study time-dependent systems explicitly.

A bonus is the extension to the arbitrary time-dependent frequency $\omega(t)$ of the Maslov phase correction [2,4–6,19,24–28] even when no explicit solutions are available (see Section 4).

In Section 5.2, we illustrate our theory by the time-dependent Mathieu profile $\omega^2(t) = a - 2q\cos 2t$, a, b const., the direct analytic treatment of which is complicated.

2. The Junker–Inomata Derivation of the Propagator

Starting with a general quadratic Lagrangian in $1 + 1$ spacetime dimensions with coordinates $\tilde{x}$ and t, Junker and Inomata derived the equation of motion [12]:

$$\ddot{\tilde{x}} + \lambda(t)\dot{\tilde{x}} + \omega^2(t)\tilde{x} = F(t)\,, \tag{11}$$

which describes a nonrelativistic particle of unit mass with dissipation $\lambda(t)$. The driving force $F(t)$ can be eliminated by subtracting a particular solution $h(t)$ of (11), $x(t) = \tilde{x}(t) - h(t)$, in terms of which (11) becomes homogeneous,

$$\ddot{x} + \lambda(t)\dot{x} + \omega^2(t)x = 0\,. \tag{12}$$

This equation can be obtained from the time-dependent generalization of (6),

$$L_{CK} = \frac{1}{2}e^{\lambda(t)}[\dot{x}^2 - \omega^2(t)x^2]\,. \tag{13}$$

The friction can be eliminated by setting $x(t) = y(t)\,e^{-\lambda(t)/2}$, which yields a harmonic oscillator with no friction, but with a shifted frequency [29–31],

$$\ddot{y} + \Omega^2(t)y = 0 \quad \text{where} \quad \Omega^2(t) = \omega^2(t) - \frac{\dot{\lambda}^2(t)}{4} - \frac{\ddot{\lambda}(t)}{2}\,. \tag{14}$$

For $\lambda(t) = \lambda_0 t$ and $\omega = \omega_0 = $ const., for example, we obtain the usual harmonic oscillator with a constant shifted frequency, $\Omega^2 = \omega_0^2 - \lambda_0^2/4 = $ const.

The frequency is in general time-dependent, though $\Omega = \Omega(t)$; therefore, (14) is a *Sturm–Liouville equation* that can be solved analytically only in exceptional cases.

Junker and Inomata [12] followed another, more subtle path. Equation (12) is a linear equation with time-dependent coefficients, the solution of which can be searched for within the ansatz (A similar transcription was used also by Rezende [28].):

$$x(t) = \rho(t)\left(Ae^{i\bar{\omega}\tau(t)} + Be^{-i\bar{\omega}\tau(t)}\right)\,, \tag{15}$$

where A, B, and $\underline{\bar{\omega}}$ are constants and $\rho(t)$ and $\tau(t)$ functions to be found. Inserting (15) into (12), putting the coefficients of the exponentials to zero, separating the real and imaginary parts, and absorbing a new integration constant into A, B provide us with the coupled system for $\rho(t)$ and $\tau(t)$,

$$\ddot{\rho} + \lambda\dot{\rho} + (\omega^2(t) - \bar{\omega}^2\dot{\tau}^2)\rho = 0, \tag{16}$$

$$\dot{\tau}(t)\,\rho^2(t)\,e^{\lambda(t)} = 1\,. \tag{17}$$

Manifestly, $\dot{\tau} > 0$. Inserting $\dot{\tau}$ into (16) then yields the *Ermakov–Milne–Pinney* (EMP) equation [32–34] with time-dependent coefficients,

$$\ddot{\rho} + \lambda\dot{\rho} + \omega^2(t)\rho = \frac{e^{-2\lambda(t)}\bar{\omega}^2}{\rho^3}\,. \tag{18}$$

We note for later use that eliminating ρ would yield instead:

$$\bar{\omega}^2 = \frac{1}{\dot{\tau}^2}\left(\omega^2(t) - \frac{1}{2}\frac{\dddot{\tau}}{\dot{\tau}} + \frac{3}{4}\left(\frac{\ddot{\tau}}{\dot{\tau}}\right)^2 - \frac{\ddot{\lambda}}{2} - \frac{\dot{\lambda}^2}{4}\right)\,. \tag{19}$$

Conversely, the constancy of the r.h.s. here can be verified using Equation (17). Equivalently, starting with the Junker–Inomata condition (10),

$$\omega^2(t) = \frac{\ddot{f}}{f} - 2\frac{\dot{f}^2}{f^2} + \frac{\lambda^2}{4} + \frac{\ddot{\lambda}}{2}.$$

(20)

To sum up, the strategy to follow is [12,35,36]:

1. to solve first the EMP Equation (18) for ρ;
2. to integrate (17),

$$\tau(t) = \int^t \frac{e^{-\lambda(u)}}{\rho^2(u)}\, du.$$

(21)

Then, the trajectory is given by (15).

Junker and Inomata showed, moreover, that substituting into (13) the new coordinates:

$$T = \frac{\tan[\bar{\omega}\,\tau(t)]}{\bar{\omega}}, \qquad X = x\,e^{\frac{\lambda(t)}{2}}\dot{\tau}(t)^{\frac{1}{2}}\sec[\bar{\omega}\,\tau(t)],$$

(22)

allows us to present the Caldirola–Kanai action as (Surface terms do not change the classical equations of motion and multiply the propagator by an unobservable phase factor, and are therefore dropped.),

$$\mathcal{A}_{CK} = \int_{t'}^{t''} L_{CK}\,dt = \int_{T'}^{T''} \frac{1}{2}\left(\frac{dX}{dT}\right)^2 dT,$$

(23)

where we recognize the *action of a free particle* of unit mass. One checks also directly that X, T satisfy the free equation, as they should. The conditions (10) are readily verified.

The coordinates X and T describe a free particle; therefore, the propagator is (5) (as anticipated by our notation). The clue of Junker and Inomata [12] is that, conversely, trading X and T in (5) for x and t allows deriving the propagator for the CK oscillator (see also [11], Section 5.1) (The extension of (24) from $0 < \bar{\omega}(\tau'' - \tau') < \pi$ to all t [2,3,6,11] is discussed in Section 4.),

$$K_{osc}(x'',t''|x',t') = \left[\frac{\bar{\omega}e^{\frac{\lambda''+\lambda'}{2}}(\dot{\tau}''\dot{\tau}')^{\frac{1}{2}}}{2\pi i\hbar\,\sin[\bar{\omega}(\tau'' - \tau')]}\right]^{\frac{1}{2}} \times$$

(24)

$$\exp\left\{\frac{i\bar{\omega}}{2\hbar\,\sin[\bar{\omega}(\tau'' - \tau')]}\left[(x''^2 e^{\lambda''}\dot{\tau}'' + x'^2 e^{\lambda'}\dot{\tau}')\cos[\bar{\omega}(\tau'' - \tau')] - 2x''x'e^{\frac{\lambda''+\lambda'}{2}}(\dot{\tau}''\dot{\tau}')^{\frac{1}{2}}\right]\right\},$$

where we used the shorthands $\lambda'' = \lambda(t'')$, $\tau'' = \tau(t'')$, etc.

This remarkable formula says that in terms of "redefined time", τ, the problem is essentially one with a constant frequency. Equation (24) is still implicit, though, as it requires solving first the coupled system (17), which we can do only in particular cases.

- When $\lambda(t) = \lambda_0 t$ where $\lambda_0 = \text{const.} \geq 0$, Equation (12) describes a time-dependent oscillator with constant friction,

$$\ddot{x} + \lambda_0\,\dot{x} + \omega^2(t)x = 0.$$

(25)

Then, setting $R(t) = \rho(t)\,e^{\lambda_0 t/2}$, Equation (17) provide us with the EMP equation for R, cf. (18),

$$\ddot{R} + \Omega^2(t)R - \frac{\bar{\omega}^2}{R^3} = 0, \quad \text{where} \quad \Omega^2(t) = \omega^2(t) - \frac{\lambda_0^2}{4};$$

(26)

- If, in addition, the *frequency* is *constant* $\omega(t) = \omega_0 = \text{const.}$, then Equation (26) is solved algebraically by:

$$\bar{\omega}^2 = \omega_0^2 - \lambda_0^2/4, \quad R = 1 \quad \Rightarrow \quad \rho(t) = e^{-\lambda_0 t/2}, \quad \tau(t) = t.$$

(27)

Thus, $x(t)$ is a linear combination of $e^{-\frac{1}{2}\lambda_0 t}\sin\bar{\omega}t$ and $e^{-\frac{1}{2}\lambda_0 t}\cos\bar{\omega}t$. The spacetime coordinate transformation of $(x,t)\to(X,T)$ in (22) simplifies to the friction-generalized form of that of Niederer [16],

$$T = \frac{\tan(\bar{\omega}t)}{\bar{\omega}}, \quad X = x\exp\left(\frac{1}{2}\lambda_0 t\right)\sec(\bar{\omega}t), \tag{28}$$

for which the general expression (24) reduces to (9) when $\lambda_0 = 0$;

- When the oscillator is turned off, $\omega_0 = 0$, but $\lambda_0 > 0$, we have motion in a dissipative medium. The coordinate transformation propagator (22) and (24) become:

$$X = \frac{2x}{1 + \exp(-\lambda_0 t)}, \quad T = \frac{2}{\lambda_0}\frac{1 - \exp(-\lambda_0 t)}{1 + \exp(-\lambda_0 t)} \tag{29}$$

and:

$$
\begin{aligned}
K_{diss}(x'',t''|x',t') &= \left[\frac{\lambda_0}{2\pi i\hbar[\exp(-\lambda_0 t') - \exp(-\lambda_0 t'')]}\right]^{\frac{1}{2}} \\
&\times \exp\left\{\frac{i\lambda_0}{2\hbar}\frac{(x'' - x')^2}{\exp(-\lambda_0 t') - \exp(-\lambda_0 t'')}\right\},
\end{aligned} \tag{30}
$$

respectively. A driving force F_0 (e.g., terrestrial gravitation) could be added and then removed by $x \to x + (F_0/\lambda_0)t$.

Further examples can be found in [13–15]. An explicitly time-dependent example is presented in Section 5.2.

3. The Eisenhart–Duval Lift

Further insight can be gained by "Eisenhart–Duval (E-D) lifting" the system to one higher dimension to what is called a "Bargmann space" [17–21]. The latter is a $d + 1 + 1$-dimensional manifold endowed with a Lorentz metric, the general form of which is:

$$g_{\mu\nu}dx^\mu dx^\nu = g_{ij}(x,t)dx^i dx^j + 2dtds - 2V(x,t)dt^2, \tag{31}$$

which carries a covariantly constant null Killing vector ∂_s. Then:

Theorem 1 ([18,20]). *Factoring out the foliation generated by ∂_s yields a nonrelativistic spacetime in $d + 1$ dimensions. Moreover, the null geodesics of the Bargmann metric $g_{\mu\nu}$ project to ordinary spacetime, consistent with Newton's equations. Conversely, if $(\gamma(t),t)$ is a solution of the nonrelativistic equations of motion, then its null lifts to Bargmann space are:*

$$\left(\gamma(t),t,s(t)\right), \quad s(t) = s_0 - \mathcal{A}(\gamma) = s_0 - \int^t L(\gamma(r),r)dr \tag{32}$$

where s_0 is an arbitrary initial value.

Let us consider, for example, a particle of unit mass with the Lagrangian of:

$$L = \frac{1}{2\alpha(t)}g_{ij}(x^k)\dot{x}^i\dot{x}^j - \beta(t)V(x^i,t), \tag{33}$$

where $g_{ij}(x^k)dx^i dx^j$ is a positive metric on a curved configuration space Q with local coordinates x^i, $i = 1,\ldots,d$. The coefficients $\alpha(t)$ and $\beta(t)$ may depend on time t, and $V(x^i,t)$ is some (possibly time-dependent) scalar potential. The associated equations of motion are:

$$\frac{d^2x^i}{dt^2} + \Gamma^i_{jk}\frac{dx^j}{dt}\frac{dx^k}{dt} - \frac{\dot{\alpha}}{\alpha}\frac{dx^i}{dt} = -\alpha\beta g^{ij}\partial_j V, \tag{34}$$

where the Γ^i_{jk} are the Christoffel symbols of the metric g_{ij}. For $d = 1$, $g_{ij} = \delta_{ij}$ and $V = \frac{1}{2}\omega^2(t)x^2$ for $\alpha = \beta = 1$, resp. for $\alpha = \beta^{-1} = e^{-\lambda(t)}$, we obtain a (possible time-dependent) 1d oscillator without, resp. with, friction, Equation (7) [7–9,29–31].

Equation (34) can also be obtained by projecting a null-geodesic of $d + 1 + 1$-dimensional Bargmann spacetime with coordinates $(x^\mu) = (x^i, t, s)$, whose metric is:

$$g_{\mu\nu}dx^\mu dx^\nu = \frac{1}{\alpha}g_{ij}dx^i dx^j + 2dtds - 2\beta V dt^2. \tag{35}$$

For $\alpha = \beta^{-1} = e^{-\lambda(t)}$, we recover (12).

Choosing $\lambda(t) = \ln m(t)$ would describe motion with a time-dependent mass $m(t)$. The friction can be removed by the conformal rescaling $x \to y = \sqrt{m}\,x$, and the null geodesics of the rescaled metric describe, consistent with (14), an oscillator with no friction, but with a time-dependent frequency, $\Omega^2 = \omega^2 - \frac{\ddot{m}}{2m} + \left(\frac{\dot{m}}{2m}\right)^2$ [37].

The friction term $-(\dot{\alpha}/\alpha)\dot{x}^i$ in (34) can be removed also by introducing a new time parameter $\tilde{t}$, defined by $d\tilde{t} = \alpha\,dt$ [21]. For $\lambda(t) = \lambda_0 t$, for example, putting $\tilde{t} = -e^{-\lambda_0 t}/\lambda_0$ eliminates the friction, but it does this at the price of obtaining a manifestly time-dependent frequency [38,39]:

$$\frac{d^2x}{d\tilde{t}^2} + \tilde{\Omega}^2(\tilde{t})x = 0, \qquad \tilde{\Omega}^2(\tilde{t}) = \frac{\omega^2}{\tilde{t}^2\lambda_0^2}. \tag{36}$$

3.1. The Junker–Inomata Ansatz as a Conformal Transformation

The approach outlined in Section 2 admits a Bargmannian interpretation. For simplicity, we only consider the frictionless case $\lambda = 0$.

Theorem 2. *The Junker–Inomata method of converting the time-dependent system into one with a constant frequency by switching from "real" to "fake time",*

$$t \to \tau(t), \qquad \xi = \dot{\tau}\,x \tag{37}$$

induces a conformal transformation between the Bargmann metrics:

$$
\begin{array}{llll}
dx^2 + 2dtds - \omega^2(t)x^2dt^2 & \quad frequency & \omega^2(t) & (38a) \\
d\xi^2 + 2d\tau d\sigma - \bar{\omega}^2\xi^2 d\tau^2, & \quad frequency & \bar{\omega} = \text{const.} & (38b)
\end{array}
$$

$$d\xi^2 + 2d\tau d\sigma - \bar{\omega}^2\xi^2 d\tau^2 = \dot{\tau}(t)\left(dx^2 + 2dtds - \omega^2(t)x^2dt^2\right). \tag{39}$$

Proof. Putting $\mu = \ln\dot{\tau}$ allows us to present the constant-frequency $\bar{\omega}$ (19) as:

$$\bar{\omega}^2 = \dot{\tau}^{-2}\left(\omega^2(t) - \tfrac{1}{2}\ddot{\mu} + \tfrac{1}{4}\dot{\mu}^2\right). \tag{40}$$

Then, with the notation $\overset{\circ}{\xi} = d\xi/d\tau$, we find,

$$\overset{\circ}{\xi}{}^2 = \dot{\tau}^{-1}\left[\dot{x}^2 + \frac{1}{4}\dot{\mu}^2x^2 - \frac{1}{2}\ddot{\mu}x^2 + \frac{d}{dt}\left(\frac{1}{2}\dot{\mu}x^2\right)\right].$$

Let us now recall that the null lift to the Bargmann space of a spacetime curve is obtained by subtracting the classical action as the vertical coordinate,

$$d\sigma = -L(\xi, \overset{\circ}{\xi}, \tau)d\tau = -\frac{1}{2}\left(\overset{\circ}{\xi}{}^2 - \bar{\omega}^2\xi^2\right)d\tau. \tag{41}$$

Setting here $\xi = x\dot{\tau}^{1/2}$ and dropping surface terms yield, using the same procedure for the time-dependent-frequency case,

$$d\sigma = ds = -\frac{1}{2}\left(\dot{x}^2 - \omega^2(t)x^2\right)dt \tag{42}$$

up to surface terms. Then, inserting all our formulae into (38a) and (38b) yields (39), as stated. In Junker–Inomata language (10), $f(t) = \tau^{1/2}\sec(\bar{\omega}\tau)$, $g(t) = (\bar{\omega})^{-1}\tan(\bar{\omega}\tau)$. $\square$

Our investigation has so far concerned classical aspects. Now, we consider what happens quantum mechanically. Restricting our attention at $d = 1$ space dimensions as before (In $d > 2$, conformal invariance requires adding a scalar curvature term to the Laplacian.), we posit that the E-D lift $\widetilde{\psi}$ of a wave function ψ is equivariant,

$$\widetilde{\psi}(x,t,s) = e^{\frac{i}{\hbar}s}\psi(x,t) \quad \Rightarrow \quad \partial_s\widetilde{\psi} = \frac{i}{\hbar}\widetilde{\psi}. \tag{43}$$

Then, the massless Klein–Gordon equation for $\widetilde{\psi}$ associated with the $1 + 1 + 1 = 3$ d Barmann metric implies the Schrödinger equation in 1+1 d,

$$\Delta_g\,\widetilde{\psi} = 0 \quad \Rightarrow \quad i\partial_t\psi = \left[-\frac{\hbar^2}{2}\Delta_x + V(x,t)\right]\psi \tag{44}$$

where Δ_g is the Laplace–Beltrami operator associated with the metric. In $d = 1$, it is of course $\Delta_x = \partial_x^2$.

A conformal diffeomorphism $(X,T,S) \to \tilde{f}(X,T,S) = (x,t,s)$ with conformal factor σ_f^2, $\tilde{f}^*g_{\mu\nu} = \sigma_f^2\,g_{\mu\nu}$, projects to a spacetime transformation $(X,T) \to f(X,T) = (x,t)$. It is implemented on a wave function lifted to the Bargmann space as:

$$\widetilde{\psi}(x,t,s) = \sigma_f^{-1/2}\widetilde{\psi}(X,T,S) \tag{45}$$

In Sections 4.2, these formulae are applied to the Niederer map (73).

3.2. The Arnold Map

The general damped harmonic oscillator with time-dependent driving force $F(t)$ in $1 + 1$ dimensions, (11),

$$\ddot{x} + \lambda\dot{x} + \omega^2(t)x = F(t), \tag{46}$$

can be solved by an *Arnold transformation* [22,23], which "straightens the trajectories" [21,29–31,40]. To this end, one introduces new coordinates,

$$T = \frac{u_1}{u_2}, \quad X = \frac{x - u_p}{u_2}, \tag{47}$$

where u_1 and u_2 are solutions of the associated homogeneous Equation (46) with $F \equiv 0$ and u_p is a particular solution of the full Equation (46). It is worth noting that (47) allows checking, independently, the Junker–Inomata criterion in (10). The initial conditions are chosen as,

$$u_1(t_0) = \dot{u}_2(t_0) = 0, \quad \dot{u}_1(t_0) = u_2(t_0) = 1, \quad u_p(t_0) = \dot{u}_p(t_0) = 0. \tag{48}$$

Then, in the new coordinates, the motion becomes free [22,23],

$$X(T) = aT + b, \quad a, b = \text{const.} \tag{49}$$

Equation (46) can be obtained by projecting a null geodesic of the Bargmann metric:

$$g_{\mu\nu}dx^\mu dx^\nu = e^{\lambda(t)}dx^2 + 2dtds - 2e^{\lambda(t)}\left(\frac{1}{2}\omega(t)^2x^2 - F(t)x\right)dt^2. \tag{50}$$

Completing (47) by:

$$S = s + e^\lambda u_2^{-1}\left(\frac{1}{2}\dot{u}_2x^2 + \dot{u}_px\right) + g(t) \quad \text{where} \quad \dot{g} = \frac{1}{2}e^\lambda\left(\dot{u}_p^2 - \omega^2u_p^2 + 2Fu_p\right) \tag{51}$$

lifts the Arnold map to Bargmann spaces, $(x, t, s) \rightarrow (X, T, S)$ (In the Junker–Inomata setting (10), $f = u_2 e^{-\lambda/2}$ and $g(t) = u_1/u_2$.),

$$g_{\mu\nu} dx^\mu dx^\nu = e^{\lambda(t)} u_2^2(t) \left(dX^2 + 2 dT dS \right). \tag{52}$$

The oscillator metric (50) is thus carried conformally to the free one, generalizing earlier results [18,19,41]. For the damped harmonic oscillator with $\lambda(t) = \lambda_0 t$ and $F(t) \equiv 0$, $u_p \equiv 0$ is a particular solution. When $\omega = \omega_0 = \text{const.}$, for example,

$$u_1 = e^{-\lambda_0 t/2} \frac{\sin \Omega_0 t}{\Omega_0}, \quad u_2 = e^{-\lambda_0 t/2} \left(\cos \Omega_0 t + \frac{\lambda_0}{2\Omega_0} \sin \Omega_0 t \right), \quad \Omega_0^2 = \omega_0^2 - \lambda_0^2/4 \tag{53}$$

are two independent solutions of the homogeneous equation with initial conditions (48) and provide us with:

$$T = \frac{\sin \Omega_0 t}{\Omega_0 (\cos \Omega_0 t + \frac{\lambda_0}{2\Omega_0} \sin \Omega_0 t)}, \tag{54}$$

$$X = \frac{e^{\lambda_0 t/2} x}{\cos \Omega_0 t + \frac{\lambda_0}{2\Omega_0} \sin \Omega_0 t}, \tag{55}$$

$$S = s - \frac{1}{2} e^{\lambda_0 t} x^2 \left(\frac{\omega_0^2}{\Omega_0} \right) \frac{\sin \Omega_0 t}{\cos \Omega_0 t + \frac{\lambda_0}{2\Omega_0} \sin \Omega_0 t}. \tag{56}$$

In the undamped case, $\lambda_0 = 0$; thus, $\Omega_0 = \omega_0$, and (56) reduces to that of Niederer [16] lifted to the Bargmann space [19,20],

$$T = \frac{\tan \omega_0 t}{\omega_0}, \quad X = \frac{x}{\cos \omega_0 t}, \quad S = s - \frac{1}{2} x^2 \omega_0 \tan \omega_0 t. \tag{57}$$

The Junker–Inomata construction in Section 2 can be viewed as a particular case of the Arnold transformation. We chose $u_p \equiv 0$ and the two independent solutions:

$$u_1 = e^{-\lambda/2} \dot{\tau}^{-1/2} \frac{\sin \bar\omega \tau}{\bar\omega}, \quad u_2 = e^{-\lambda/2} \dot{\tau}^{-1/2} \cos \bar\omega \tau. \tag{58}$$

The initial conditions (48) at $t_0 = 0$ imply $\tau(0) = \dot\rho(0) = 0$, $\rho(0) = \dot\tau(0) = 1$. Then, spelling out (51),

$$S = s - \frac{1}{2} e^\lambda \left(\bar\omega \dot\tau \tan \bar\omega \tau + \frac{1}{2}\dot\lambda + \frac{1}{2}\frac{\ddot\tau}{\dot\tau} \right) x^2 \tag{59}$$

completes the lift of (22) to Bargmann spaces. In conclusion, the one-dimensional damped harmonic oscillator is described by the conformally flat Bargmann metric,

$$g_{\mu\nu} dx^\mu dx^\nu = \frac{\cos^2 \bar\omega \tau}{\dot\tau} \left(dX^2 + 2 dT dS \right). \tag{60}$$

The metric (60) is manifestly conformally flat; therefore, its geodesics are those of the free metric, $X(T) = aT + b$. Then, using (47) with (58) yields:

$$x(t) = e^{-\lambda(t)/2} \dot{\tau}^{-1/2}(t) \left(a \frac{\sin[\bar\omega \tau(t)]}{\bar\omega} + b \cos[\bar\omega \tau(t)] \right). \tag{61}$$

The bracketed quantity here describes a constant-frequency oscillator with "time" $\tau(t)$. The original position, x, obtains a time-dependent "conformal" scale factor.

4. The Maslov Correction

As mentioned before, the semiclassical formula (9) is correct only in the first oscillator half-period, $0 < t'' - t' < \pi/\Omega_0$. Its extension for all t involves the *Maslov correction*. In the constant-frequency case with no friction, for example, assuming that $\Omega_0(t'' - t'')/\pi$ is not an integer, we have [2,3,6],

$$K^{ext}(x'', t''|x', t') = \left[\frac{\Omega_0}{2\pi\hbar\,|\sin\Omega_0(t''-t')|}\right]^{\frac{1}{2}} \times e^{-i\frac{\pi}{4}(1+2\ell)} \tag{62}$$

$$\exp\left\{\frac{i\Omega_0}{2\hbar\,\sin\Omega_0(t''-t')}\left[(x''^2 + x'^2)\cos\Omega_0(t''-t') - 2x''x'\right]\right\},$$

where the integer:

$$\ell = \mathrm{Ent}\left[\frac{\Omega_0(t''-t')}{\pi}\right] \tag{63}$$

is called the *Maslov index* ($\mathrm{Ent}[x]$ is the integer part of x.). ℓ counts the completed half-periods and is related also to the *Morse index*, which counts the negative modes of $\partial^2 A/\partial x'\partial x''$ [4,5].

Now, we generalize (62) to the *time-dependent* frequency:

Theorem 3. *In terms of $\bar\omega$ and τ introduced in Section 2,*

- *Outside caustics, i.e., for $\bar\omega(\tau'' - \tau') \neq \pi\ell$, the propagator for the harmonic oscillator with the time-dependent frequency and friction is:*

$$K^{ext}(x'', t''|x', t') = \left[\frac{\bar\omega e^{\frac{\lambda''+\lambda'}{2}}(\dot\tau''\dot\tau')^{\frac{1}{2}}}{2\pi\hbar|\sin\bar\omega(\tau''-\tau')|}\right]^{1/2} \exp\left\{-\frac{i\pi}{2}\left(\frac{1}{2} + \mathrm{Ent}\left[\frac{\bar\omega(\tau''-\tau')}{\pi}\right]\right)\right\} \tag{64}$$

$$\times \exp\left\{\frac{i\bar\omega}{2\hbar\sin\bar\omega(\tau''-\tau')}\left[(x''^2 e^{\lambda''}\dot\tau'' + x'^2 e^{\lambda'}\dot\tau')\cos[\bar\omega(\tau''-\tau')] - 2x''x'e^{\frac{\lambda''+\lambda'}{2}}(\dot\tau''\dot\tau')^{\frac{1}{2}}\right]\right\} \quad ;$$

- *At caustics, i.e., for:*

$$\bar\omega(\tau'' - \tau') = \pi\ell, \quad \ell = 0, \pm1, \ldots \tag{65}$$

we have instead [3,6],

$$K^{ext}\left(x'', x', |\tau'' - \tau' = \frac{\pi}{\bar\omega}\ell\right) = \left[e^{\frac{\lambda''+\lambda'}{2}}(\dot\tau''\dot\tau')^{\frac{1}{2}}\right]^{1/2} \tag{66}$$

$$\times \exp\left(-\frac{i\pi\ell}{2}\right)\delta\left(x'\exp(\lambda'/2)\dot\tau'^{1/2} - (-1)^k x''\exp(\lambda''/2)\dot\tau''^{1/2}\right).$$

Proof. In terms of the redefined coordinates:

$$\tau = \tau(t) \quad \text{and} \quad \xi = x\exp\left[\frac{\lambda(t)}{2}\right]\dot\tau^{1/2}(t), \tag{67}$$

cf. (37), and using the notation $\{\overset{\circ}{\cdot}\} = d/d\tau$, the time-dependent oscillator Equation (12) is taken into:

$$\overset{\circ\circ}{\xi} + \bar\omega^2\xi = 0, \quad \text{where} \quad \bar\omega^2 = \frac{1}{\dot\tau^2}\left(\omega^2(t) - \frac{1}{2}\frac{\dddot\tau}{\dot\tau} + \frac{3}{4}\left(\frac{\ddot\tau}{\dot\tau}\right)^2 - \frac{\ddot\lambda}{2} - \frac{\dot\lambda^2}{4}\right). \tag{68}$$

Thus, the problem is reduced to one with a *time-independent* frequency, $\bar\omega$ in (19) (We record for the sake of later investigations that (turning off λ) (68) can be presented as:

$$\omega^2(t) - \dot\tau^2\,\bar\omega^2 = \tfrac{1}{2}S(\tau) \tag{69}$$

where $S(\tau) = \frac{\dddot{\tau}}{\dot{\tau}} - \frac{3}{2}\left(\frac{\ddot{\tau}}{\dot{\tau}}\right)^2$ is the *Schwarzian derivative* of τ [42]). □

Let us now recall Formula (19) of Junker and Inomata in [12], which tells us how propagators behave under the coordinate transformation $(\xi, \tau) \longleftrightarrow (x, t)$:

$$K_2(x'', t''|x', t') = \left[\left(\frac{\partial \xi'}{\partial x'}\right)\left(\frac{\partial \xi''}{\partial x''}\right)\right]^{\frac{1}{2}} K_1(\xi'', \tau''|\xi', \tau'). \tag{70}$$

Here, $K_2 = K^{ext}$ is the propagator of an oscillator with a time-dependent frequency and friction, $\omega(t)$ and $\lambda(t)$, respectively—the one we are trying to find. K_1 is in turn the Maslov-extended propagator of an oscillator with no friction and a constant frequency, as in (62). Then, the propagator for the *harmonic oscillator with a time-dependent frequency and friction*, Equation (64), is obtained using (67).

Notice that (64) is *regular* at the points $r_k \in J_k$ where $\sin = \pm 1$. However, at caustics, $\tau'' - \tau' = (\pi/\bar{\omega})\ell$, K^{ext} diverges, and we have instead (66).

Henceforth, we limit our investigations to $\lambda = 0$.

4.1. Properties of the Niederer Map

More insight is gained from the perspective of the generalized Niederer map (22). We first study their properties in some detail. For simplicity, we chose, in the rest of this section, $x' = t' = 0$ and $x'' \equiv x$ and $t'' \equiv t$.

We start with the observation that the Niederer map (22) becomes singular where the cosine vanishes, i.e., where:

$$\cos[\bar{\omega}\tau(r_k)] = 0, \quad \text{i.e.} \quad \tau(r_k) = (k + \tfrac{1}{2})\frac{\pi}{\bar{\omega}}, \quad k = 0, \pm 1, \ldots \tag{71}$$

$r_k < r_{k+1}$ because $\tau(t)$ is an increasing function by (21). Moreover, each interval:

$$I_k = [r_k, r_{k+1}], \quad k = 0, \pm 1, \ldots \tag{72}$$

is mapped by (22) onto the full range $-\infty < T < \infty$. Therefore, the inverse mapping is *multivalued*, labeled by integers k,

$$N_k : T \to t = \frac{\arctan_k \bar{\omega}T}{\bar{\omega}}, \quad X \to x = \frac{X}{\sqrt{1 + \bar{\omega}^2 T^2}}, \tag{73}$$

where $\arctan_k(\,\cdot\,) = \arctan_0(\,\cdot\,) + k\pi$ with $\arctan_0(\,\cdot\,)$ the principal determination, i.e., in $(-\pi/2, \pi/2)$.

Then, $\lim_{t \to r_k-} \tan t = \infty$ and $\lim_{t \to r_k+} \tan t = -\infty$ imply that:

$$\lim_{T \to \infty} N_k(T) = r_{k+1} = \lim_{T \to -\infty} N_{k+1}(T). \tag{74}$$

Therefore, the intervals I_k and I_{k+1} are joined at r_{k+1}, and the I_k form a partition of the time axis, $\{-\infty < t < \infty\} = \cup_k I_k$.

Returning to (64) (which is (62) with $\Omega_0 \Rightarrow \bar{\omega}$, $t \Rightarrow \tau$), we then observe that, whereas the propagator is regular at r_k, it diverges at caustics,

$$\sin[\bar{\omega}\tau(t_\ell)] = 0 \quad \text{i.e.,} \quad \tau(t_\ell) = \frac{\pi}{\bar{\omega}}\ell, \quad \ell = 0, \pm 1, \ldots, \tag{75}$$

cf. (65). Thus, $t_\ell \leq t_{\ell+1}$, and:

$$N_k(-\infty) = r_k, \quad N_k(T = 0) = t_{k+1}, \quad N_k(+\infty) = r_{k+1}. \tag{76}$$

Thus, N_k maps the full T-line into I_k with t_k an internal point. Conversely, r_k is an *internal point* of J_k. The intervals $J_\ell = [t_\ell, t_{\ell+1}]$ cover again the time axis, $\cup_\ell J_\ell = \{-\infty < t < \infty\}$.

By (61) the classical trajectories are regular at $t = r_k$. Moreover, for arbitrary initial velocities,

$$\sqrt{\dot{\tau}(t_{\ell+1})}\, x(t_{\ell+1}) = -\sqrt{\dot{\tau}(t_\ell)}\, x(t_\ell) \tag{77}$$

implying that after a half-period $\bar{\omega}\tau \to \bar{\omega}\tau + \pi$, all classical motions are focused at the same point. The two entangled sets of intervals are shown in Figure 1.

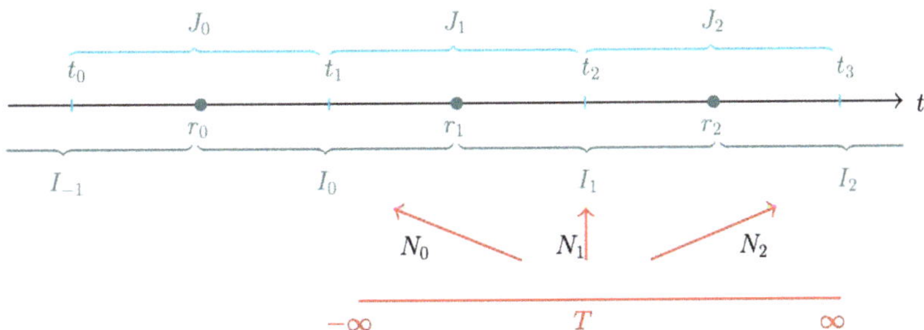

Figure 1. The generalized Niederer map (22) maps each interval $I_k = (r_k, r_{k+1})$ onto the entire real line $-\infty < T < \infty$. Its inverse mapping is therefore multivalued, labeled by an integer k. The classical motions and the propagator are both regular at the separation points r_k. All classical trajectories are focused at the caustic points $t_\cdot$, where the propagator diverges.

The Niederer map (57) "E-D lifts" to the Bargmann space.

Theorem 4. *The E-D lift of the inverse of the Niederer map* (57)*, which we shall denote by* $\widetilde{N}_k : (X, T, S) \to (x, t, s)$ $(t \in I_k)$*, is:*

$$t = \frac{\arctan_k \bar{\omega} T}{\bar{\omega}}, \quad x = \frac{X}{\sqrt{1 + \bar{\omega}^2 T^2}}, \quad s = S + \frac{X^2}{2}\frac{\bar{\omega}^2 T}{1 + \bar{\omega}^2 T^2}. \tag{78}$$

Proof. These formulae follow at once by inverting (57), at once with the cast $\omega_0 \Rightarrow \bar{\omega}$, $t \Rightarrow \tau$. Alternatively, it could also be proven as for Theorem 2.

For each integer k (78) maps the real line $-\infty < T < \infty$ into the "open strip" [19] $[r_k, r_{k+1}] \times \mathbb{R}^2 \equiv I_k \times \mathbb{R}^2$ with r_k defined in (71). Their union covers the entire Bargmann manifold of the oscillator. $\square$

Now, we pull back the free dynamics by the multivalued inverse (78). We put $\bar{\omega} = 1$ for simplicity. The free motion with initial condition $X(0) = 0$,

$$X(T) = aT, \quad S(T) = S_0 - \frac{a^2}{2}T, \tag{79}$$

E-D lifts by (78) to:

$$x(t) = a \sin t \qquad s(t) = S_0 - \frac{a^2}{4}\sin 2t, \tag{80}$$

consistent with $s(t) = s_0 - \mathcal{A}_{osc}$, as can be checked directly. Note that the s coordinate oscillates with a doubled frequency.

- At $t = r_k = (\frac{1}{2} + k)\pi$ (where the Niederer maps are joined), we have $\lim_{t \to r_k} x(t) = (-1)^{k+1}a$, $\lim_{t \to r_k} s(t) = S_0$. Thus, the pull backs of the Bargmann lifts of free motions are glued to smooth curves;
- Similarly, at t caustics $t = t_\ell = \pi\ell$, we infer from (80) that for all initial velocities a and for all ℓ $\lim_{t \to t_\ell} x(t) = 0$, $\lim_{t \to t_\ell} s(t) = S_0$. Thus, the lifts are again smooth at t_ℓ, and after each half-period, all motions are focused above the initial position $(x(0) = 0, s(0) = S_0)$.

4.2. The Propagator by the Niederer Map

Now, we turn to quantum dynamics. Our starting point is the free propagator (5), which (as mentioned before) is valid only for $0 < T'' - T'$. Its extension to all T involves the *sign* of $(T'' - T')$ [19].

Let us explain this subtle point in some detail. First of all, we notice that the usual expression (5) involves a square root, which is double-valued, obliging us to *choose* one of its branches. Which one we choose is irrelevant: it is a mere gauge choice. However, once we do choose one, we must stick to our choice. Take, for example, the one for which $\sqrt{-i} = e^{-i\pi/4}$, then the prefactor in (5) is:

$$\left[\frac{1}{2\pi i\hbar(T'' - T')}\right]^{1/2} = e^{-i\pi/4}\left[\frac{1}{2\pi\hbar|T'' - T'|}\right]^{1/2}.$$

Let us now consider what happens when $T'' - T'$ changes sign. Then, the prefactor becomes multiplied by $\sqrt{-1}$ so it becomes, *for the same choice of the square root,*

$$e^{i\pi/2}e^{-i\pi/4}\left[\frac{1}{2\pi\hbar|T'' - T'|}\right]^{1/2} = e^{+i\pi/4}\left[\frac{1}{2\pi\hbar|T'' - T'|}\right]^{1/2}. \tag{81}$$

In conclusion, the formula valid for all T is,

$$K_{free}(X'', T''|X', T') = e^{-i\frac{\pi}{4}\operatorname{sign}(T''-T')}\left[\frac{1}{2\pi\hbar|T'' - T'|}\right]^{1/2}\exp\left\{\frac{i}{\hbar}\tilde{A}_{free}\right\}, \tag{82}$$

where:

$$\tilde{A}_{free} = \frac{(X'' - X')^2}{2(T'' - T')} \tag{83}$$

is the free action calculated along the classical trajectory. Let us underline that (82) already involves a "Maslov jump" $e^{-i\pi/2}$, which, for a free particle, happens at $T = 0$. For $T'' - T' = 0$, we have $K_{free} = \delta(X'' - X')$.

Accordingly, the wave function $\Psi \equiv \Psi_{free}$ of a free particle is, by (1),

$$\Psi(X'', T'') = e^{-i\frac{\pi}{4}\operatorname{sign}(T''-T')}\left[\frac{1}{2\pi\hbar|T'' - T'|}\right]^{1/2}\int_{\mathbb{R}}\exp\left\{\frac{i}{\hbar}\tilde{A}_{free}\right\}\Psi(X', T')dX'. \tag{84}$$

Now, we pull back the free dynamics using the multivalued inverse Niederer map. It is sufficient to consider the constant-frequency case $\bar{\omega} = \text{const.}$ and to denote time by t. Let t belong to the range of N_k in (73), $t \in I_k = [r_k, r_{k+1}] = N_k(\{-\infty < T < \infty\})$. Then, applying the general formulae in Section 3.1 yields [19],

$$\begin{aligned}
\tilde{\psi}(x'', t'', s'') &= \cos^{-1/2}[\bar{\omega}(t'' - t')]\tilde{\Psi}(X'', T'', S'') = e^{-\frac{i\pi}{4}\operatorname{sign}\left(\frac{\tan\bar{\omega}(t''-t'')}{\bar{\omega}}\right)} \times \\
&\quad \cos^{-1/2}[\bar{\omega}(t'' - t')]\exp\left(\frac{i}{\hbar}s''\right)\exp\left(-\frac{i}{\hbar}(\tfrac{1}{2}\bar{\omega}x''^2\tan[\bar{\omega}(t'' - t'')])\right) \\
&\quad \sqrt{\frac{|\bar{\omega}|}{2\pi\hbar|\tan[\bar{\omega}(t'' - t')]|}}\int_{\mathbb{R}}\exp\left\{\frac{i}{\hbar}\frac{\bar{\omega}|\frac{x''}{\cos[\bar{\omega}(t''-t')]} - x'|^2}{2\tan[\bar{\omega}(t'' - t')]}\right\}\psi(x', t')dx'.
\end{aligned}$$

However, the second exponential in the middle line combines with the integrand in the braces in the last line to yield *the action calculated along the classical oscillator trajectory,*

$$\tilde{A}_{osc} = \frac{\bar{\omega}}{2\sin\bar{\omega}(t'' - t')}\left((x''^2 + x'^2)\cos\bar{\omega}(t'' - t') - 2x''x'\right). \tag{85}$$

Thus, using the equivariance, we end up with,

$$\psi_{osc}(x'',t'') \;=\; \cos^{-1/2}[\bar{\omega}(t''-t')]\,\exp\!\left[-\frac{i\pi}{4}\,\mathrm{sign}\!\left(\frac{\tan[\bar{\omega}(t''-t')]}{\bar{\omega}}\right)\right]\times \tag{86}$$

$$\sqrt{\frac{|\bar{\omega}|}{2\pi\hbar|\tan[\bar{\omega}(t''-t')]|}}\int_{\mathbb{R}}\exp\!\left\{\frac{i}{\hbar}\bar{A}_{osc}\right\}\psi_{osc}(x',t')\,dx\,.$$

Now, we recover the Maslov jump, which comes from the first line here. For simplicity, we consider again $t'=0$, $x'=0$ and denote $t''=t$, $x''=x$.

Firstly, we observe that the conformal factor $\cos\bar{\omega}t$ has a constant sign in the domain I_k and changes sign at the end points. In fact,

$$\cos\bar{\omega}t = (-1)^{k+1}|\cos\bar{\omega}t| \;\;\Rightarrow\;\; \cos^{-1/2}(\bar{\omega}t) = e^{-i\frac{\pi}{2}(k+1)}|\cos\bar{\omega}t|^{-1/2}. \tag{87}$$

The cosine enters into the van Vleck factor, while the phase combines with $\exp\!\left[-\frac{i\pi}{4}\,\mathrm{sign}(\frac{\tan\bar{\omega}t}{\bar{\omega}})\right]$. Recall now that $t_{k+1} = N_k(T=0)$ divides I_k into two pieces, $I_k = [r_k, t_{k+1}]\cup[t_{k+1}, r_{k+1}]$, cf. Figure 1. However, t_{k+1} is precisely where the tangent changes sign: this term contributes to the phase in $[r_k, t_{k+1}]$ $-\pi/4$ and $+\pi/4$ in $[t_{k+1}, r_{k+1}]$. Combining the two shifts, we end up with the phase:

$$\begin{array}{ll}-\frac{\pi}{4}(1+2\ell) & \text{for}\quad r_k<t<t_{k+1}\\[4pt]-\frac{\pi}{4}(1+2(\ell+1)) & \text{for}\quad t_{k+1}<t<r_{k+1}\end{array}\qquad\text{where}\quad \ell=\mathrm{Ent}\!\left[\frac{\bar{\omega}\tau}{\pi}\right]=k+1 \tag{88}$$

which is the Maslov jump at t_ℓ.

Intuitively, the multivalued N_k "exports" to the oscillator at $t_{\ell+1}$ the phase jump of the free propagator at $T=0$. Crossing from J_ℓ to $J_{\ell+1}$ shifts the index ℓ by one.

5. Probability Density and Phase of the Propagator: A Pictorial View

5.1. For a Constant Frequency

We assume first that the frequency is constant. We split the propagator $K(x,t)\equiv K(x,t|0,0)$ in (62) as,

$$K(x,t) = |K(x,t)|\,P(t), \qquad P(t) = e^{i(phase)}. \tag{89}$$

The *probability density*,

$$|K(x,t)|^2 = \frac{\Omega_0}{2\pi\hbar\,|\sin\Omega_0 t|} \tag{90}$$

viewed as a surface above the $x-t$ plane, diverges at $t=t_\ell=\pi\ell$, $\ell=0,\pm1,\dots$.

Representing the *phase of the propagator* would require four dimensions, though. However, recall that that the dominant contribution to the path integral should come from where the phase is stationary [1], i.e., from the neighborhood of classical paths $\bar{x}(t)$, distinguished by the vanishing of the first variation, $\delta A_{\bar{x}}=0$. Therefore, we shall study the evolution of the phase along classical paths $\bar{x}(t)$ for which (61) yields, for $\hbar=\bar{\omega}=1$ and $a\in\mathbb{R}$, $b=0$,

$$\bar{x}_a(t) = a\sin t \quad\text{and}\quad P_a(t) = \exp\!\left\{-\frac{i\pi}{4}[1-\frac{a^2}{\pi}\sin 2t] - \frac{i\pi}{2}\ell\right\}, \tag{91}$$

as depicted in Figure 2.

An intuitive understanding comes by noting that when $t\neq\pi\ell=t_\ell$, then different initial velocities a yield classical paths $\bar{x}_a(t)$ with different end points, and thus contribute to different propagators. However, approaching from the left ℓ-times a half period, $t\to$

$(\pi\ell)-$, all classical paths become focused at the same end point ($x = 0$ for our choice) and for all a,

$$P_a(t \to \pi\ell-) = e^{-i\frac{\pi}{4}(1+2\ell)} \equiv P_\ell. \tag{92}$$

which is precisely the Maslov phase. Thus, *all classical paths contribute equally*, by P_ℓ, and to the *same* propagator. Comparing with the right-limit,

$$P_a(t \to \pi\ell+) = e^{-i\frac{\pi}{4}(1+2(\ell+1))} = P_{\ell+1} = e^{-\frac{i\pi}{2}} P_\ell. \tag{93}$$

the Maslov jump is recovered. Choosing instead $y \neq 0$, there will be no classical path from $(0,0)$ to $(y, \pi\ell)$, and thus no contribution to the path integral.

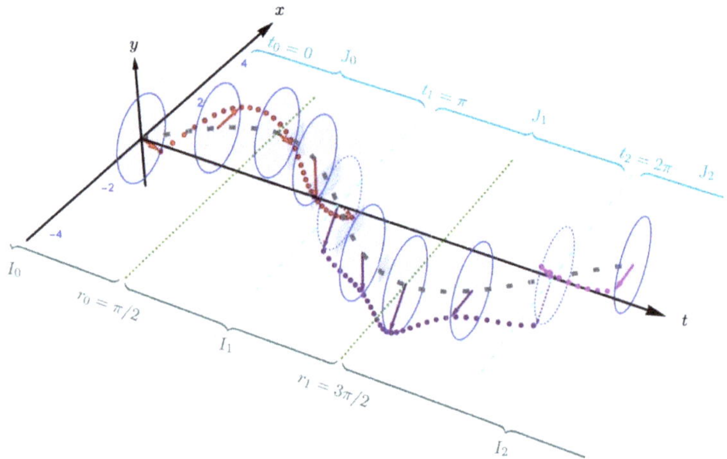

Figure 2. The phase factor $\mathbf{P}(t)$ of the propagator in (89) lies on the unit circle of the complex plane plotted vertically along a classical path $\bar{\gamma}(t)$. The orientation is positive if it is clockwise when seen from $t = +\infty$. In the time interval J_ℓ labeled by the Maslov index $\ell = \mathrm{Ent}[t/\pi]$, the factor $\mathbf{P}(t)$ precesses around $\mathbf{P}_\ell = \exp[-i\frac{\pi}{4}(1 + 2\ell)]$ with double frequency w.r.t. the classical path, $\bar{\gamma}(t)$. Arriving at a caustic, the phase jumps by $(-\pi/2)$ (**red** becoming **purple**) and then continues until the next caustic when it jumps again (and becomes **magenta**), and so on.

To conclude this section, we just mention with that the extended Feynman method [6] with the cast $\bar{\omega}$ = constant frequency and τ = "fake time" would lead also to (64) and (66) with the integer ℓ counting the number of negative eigenvalues (Morse index) of the Hessian [2,4,5,24].

5.2. A Time-Dependent Example: The Mathieu Equation

The combined Junker–Inomata–Arnold method allows us to go beyond the constant-frequency case, as illustrated here for no friction or driving force, $\lambda = F \equiv 0$, but with explicitly time-dependent frequency. For $\Omega^2(t) = a - 2q\cos 2t$, for example, (14) becomes the Mathieu equation,

$$\ddot{x} + (a - 2q\cos 2t)x = 0. \tag{94}$$

This equation can be solved either analytically using Mathieu functions [43], or numerically, providing us for $a = 2$ and $q = 1$ (for which odd Mathieu functions are real) with the dotted curve (in **red**), shown in Figure 3.

Alternatively, we can use the Junker–Inomata–Arnold transformation (47) [22,23,40]. We first achieve $\bar{\omega} = 1$ by a redefinition, $\tau \to \tau' = \bar{\omega}\tau$. Inserting Ansatz (15) into (94) yields the pair of coupled Equations (16) and (17). We chose $u_p = 0$ and two independent solutions $u_1(t)$ and $u_2(t)$, (58), with initial conditions (48) with $t_0 = 0$, i.e., $\tau(0) = \dot{\rho}(0) =$

0, $\rho(0) = \dot{\tau}(0) = 1$, which fix the integration constant, $C = \rho^2(0)\dot{\tau}(0) = 1$. Then, consistent with the general theory outlined above, the Arnold map (47) lifted to the Bargmann space becomes (22), completed with (59) with $\lambda = 0$.

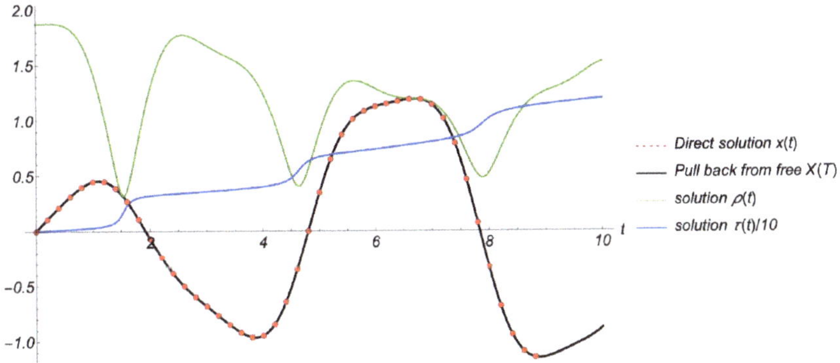

Figure 3. The analytic solution of the Mathieu equation with $a = 2, q = 1$ for $x(t)$ (**dotted** in **red**) lies on the **black** curve obtained by (15) from combining the numerically obtained $\rho(t)$ (in **green**) and $\tau(t)$ (in **blue**), which are solutions of the pair (18)–(21). The **black** curve is also obtained by pulling back the free solution (49) by the inverse Niederer map (73).

Equation (17) is solved by following the strategy outlined in Section 2. Carrying out those steps numerically provides us with Figure 3.

From the general formula (24), we deduce, for our choice $x'' = x, t'' = t, x' = t' = 0$, that the probability density (The wave function is multiplied by the square root of the conformal factor, cf. (39).).

$$|K(x,t)|^2 = \frac{\sqrt{\dot{t}}}{2\pi\hbar|\sin\tau(t)|} \, , \tag{95}$$

happens, not depending on the position, and can therefore be plotted as in Figure 4.

The propagator K and hence the probability density (95) diverge at t_i, which are roughly $t_1 \approx 1.92$, $t_2 \approx 4.80$, $t_3 \approx 7.83$. The classical motions are regular at the caustics, $\bar{x}(t_\ell) \propto \rho(t_\ell) \approx 0$; see Section 4. The domains $I_k = [r_{k-1}, r_k]$ of the inverse Niederer map are shown in Figure 4. Approximately, $r_1 \approx 1.52$, $r_2 \approx 4.49$, $r_3 \approx 6.75$, $r_4 \approx 8.44$. The evolution of the phase factor along the classical path is depicted in Figure 5.

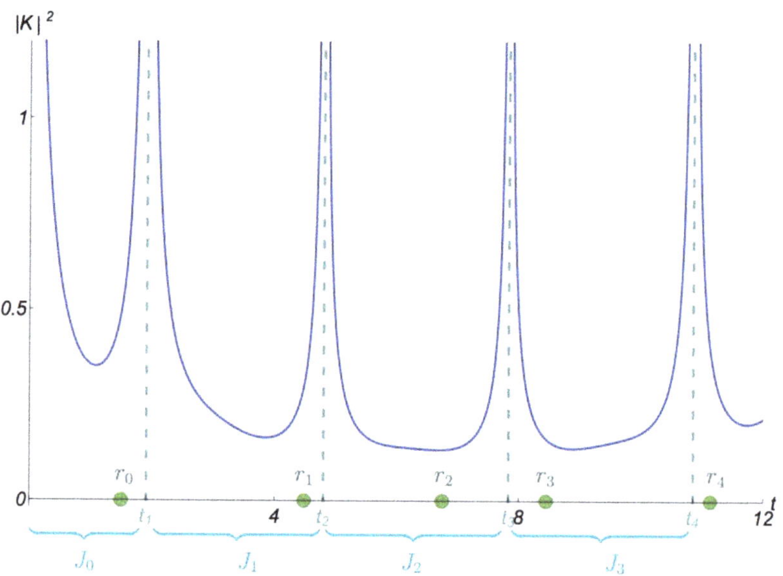

Figure 4. The probability density $|K(x,t)|^2$ (95) does not depend on x and is regular in each interval I_k between the adjacent points t_k (75), where it diverges. The r_k that determines the domains I_k of the generalized Niederer map (22) lies between the t_k and conversely.

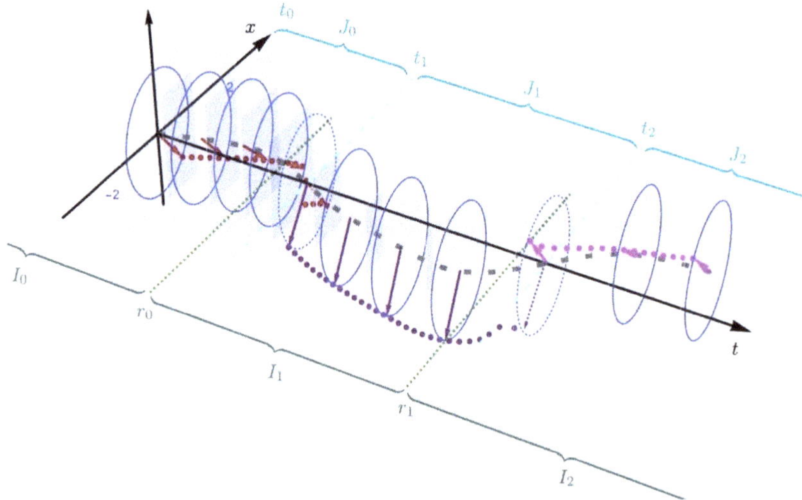

Figure 5. For $0 < t < t_1$, the Mathieu phase factor $\mathbf{P}(t)$ plotted along a classical path $\bar{\gamma}(t) = (\bar{x}(t), t)$ precesses around $e^{-i\pi/4}$. Arriving at the caustic point $\tau(t_1) = \pi$, its phase jumps by $(-\pi/2)$, then oscillates around $e^{-3i\pi/4}$ until $\tau(t_2) = 2\pi$, then jumps again, and so on.

6. Conclusions

The Junker–Inomata–Arnold approach yields (in principle) the exact propagator for any quadratic system by switching from a *time-dependent* to a *constant frequency* and redefined time,

$$\omega(t) \;\rightarrow\; \bar{\omega} = \text{const.} \quad \text{and} \quad t \;\rightarrow\; \text{``fake time''} \;\tau. \tag{96}$$

The propagator (64)–(66) is then derived from the result known for the constant frequency. A straightforward consequence is the Maslov jump for arbitrary time-dependent frequency $\omega(t)$: everything depends only on the product $\bar{\omega}\,\tau$.

By switching from t to τ, the Sturm–Liouville-type difficulty is not eliminated, but only transferred to that of finding $\tau = \tau(t)$ following the procedure outlined in Section 2. We have to first solve EMP Equation (18) for $\rho(t)$ (which is nonlinear and has time-dependent coefficients) and then integrate ρ^{-2}; see (21). Although this is as difficult to solve as solving the Sturm–Liouville equation, it provides us with theoretical insights.

When no analytic solution is available, we can resort to numerical calculations.

The Junker–Inomata approach of Section 2 is interpreted as a Bargmann-conformal transformation between time-dependent and constant frequency metrics; see Equation (39).

Alternatively, the damped oscillator can be converted to a free system by the generalized Niederer map (22), whose Eisenhart–Duval lift (47)–(51) carries the conformally flat oscillator metric (60) to the flat Minkowski space.

Two sets of points play a distinguished role in our investigations: the r_k in (71) and the t_ℓ in (75). The r_k divides the time axis into domains I_k of the (generalized) Niederer map (22). Both classical motions and quantum propagators are *regular* at r_k, where these intervals are joined. The t_ℓ are in turn the caustic points where all *classical trajectories are focused*, and the *quantum* propagator becomes *singular*.

While the "Maslov phase jump" at caustics is well established when the frequency is constant, $\omega = \omega_0 = \text{const.}$, its extension to the time-dependent case $\omega = \omega(t)$ is more subtle. In fact, the proofs we are aware of [25–28] use sophisticated mathematics, or a lengthy direct calculation of the propagator [44]. A bonus from the Junker–Inomata transcription (10) we followed here is to provide us with a straightforward extension valid to an arbitrary $\omega(t)$. Caustics arise when (65) holds, and then, the phase jump is given by (88).

The subtle point mentioned above comes from the standard (but somewhat sloppy) expression (5), which requires choosing a branch of the double-valued square root function. Once this is done, the sign change of $T'' - T'$ induces a phase jump $\pi/2$. Our "innocent-looking" factor *is* in fact the Maslov jump for a free particle at $T = 0$ (obscured when one considers the propagator for $T > 0$ only). Moreover, it then becomes the key tool for the oscillator: intuitively, the multivalued inverse Niederer map repeats, again and again, the same jump. The details are discussed in Section 4.

The transformation (10) is related to the *nonrelativistic "Schrödinger" conformal symmetries* of a free nonrelativistic particle [45–47], later extended to the oscillator [16] and an inverse square potential [18]. These results can in fact be derived using a time-dependent conformal transformation of the type (10) [19,42].

The above results are readily generalized to higher dimensions. For example, the oscillator frequency can be time-dependent, uniform electric and magnetic fields, and a curl-free "Aharonov–Bohm" potential (a vortex line [49]) can also be added [41]. Further generalization involves a Dirac monopole [50].

Alternative ways to relate free and harmonically trapped motions are studied, e.g., in [51–54]. Motions with the Mathieu profile were considered also in [55].

Author Contributions: All authors have contributed equally and substantially to this paper, including its conception, development, preparation, writing and editing. All authors have read and agreed to publish the manuscript in its present form.

Funding: This research was funded by the National Natural Science Foundation of China (grant number 11975320).

Institutional Review Board Statement: Exclude this statement since the study did not involve humans or animals.

Informed Consent Statement: Exclude this statement since the study did not involve humans.

Data Availability Statement: Exclude this statement since the study did not report any data.

Acknowledgments: This work was partially supported by the National Natural Science Foundation of China (Grant No. 11975320).

Conflicts of Interest: The authors declare no conflict of interest.

References

1. Feynman, R.P.; Hibbs, A.R. *Quantum Mechanics and Path Integrals*; McGraw-Hill: New York, NY, USA, 1965.
2. Schulman, L. *Techniques and Applications of Path Integration*; Wiley: New York, NY, USA, 1981.
3. Khandekar, D.C.; Lawande, S.V.; Bhagwat, K.V. *Path-Integral Methods and Their Applications*, 1st ed.; World Scientific: Singapore, 1993.
4. DeWitt-Morette, C. The Semiclassical Expansion. *Ann. Phys.* **1976**, *97*, 367–399; Erratum: *Ann. Phys.* **1976**, *101*, 682. [CrossRef]
5. Levit, S.; Smilansky, U. A New Approach to Gaussian Path Integrals and the Evaluation of the Semiclassical Propagator. *Ann. Phys.* **1977**, *103*, 198. [CrossRef]
6. Horvathy, P.A. Extended Feynman Formula for Harmonic Oscillator. *Int. J. Theor. Phys.* **1979**, *18*, 245. [CrossRef]
7. Caldirola, P. Forze non-conservative nella meccanica quantistica. *Nuovo Cim.* **1941**, *18*, 393. [CrossRef]
8. Kanai, E. On the Quantization of the Dissipative Systems. *Prog. Theor. Phys.* **1948**, *3*, 440. [CrossRef]
9. Dekker, H. Classical and quantum mechanics of the damped harmonic oscillator. *Phys. Rep.* **1981**, *80*, 1–112. [CrossRef]
10. Khandekar, D.C.; Lawande, S.V. Feynman Path Integrals: Some Exact Results and Applications. *Phys. Rep.* **1986**, *137*, 115–229. [CrossRef]
11. Um, C.I.; Yeon, K.H.; George, T.F. The Quantum damped harmonic oscillator. *Phys. Rep.* **2002**, *362*, 63–192. [CrossRef]
12. Junker, G.; Inomata, A. Transformation of the free propagator to the quadratic propagator. *Phys. Lett. A* **1985**, *110*, 195–198. [CrossRef]
13. Cai, P.Y.; Inomata, A.; Wang, P. Jackiw Transformation in Path Integrals. *Phys. Lett. A* **1982**, *91*, 331–334. [CrossRef]
14. Cai, P.Y.; Cai, J.M.; Inomata, A. A time-dependent conformal transformation in Feynman's path integral. In *Path integrals from meV to MeV*; World Scientific: Teaneck, NJ, USA, 1989; pp. 279–290.
15. Inomata, A. Time-dependent conformal transformation in quantum mechanics. In Proceedings of the ISATQP-Shanxi 1992, Taiyuan, China, 12–16 June 1992; Liang, J.Q., Wang, M., Qiao, S.N., Su, D.C., Eds.; Science Press: Beijing, China, 1993; pp. 75–82.
16. Niederer, U. The maximal kinematical invariance group of the harmonic oscillator. *Helv. Phys. Acta* **1973**, *46*, 191–200.
17. Eisenhart, L.P. Dynamical trajectories and geodesics. *Ann. Math.* **1928**, *30*, 591–606. [CrossRef]
18. Duval, C.; Burdet, G.; Kunzle, H.P.; Perrin, M. Bargmann Structures and Newton-cartan Theory. *Phys. Rev. D* **1985**, *31*, 1841–1853. doi:10.1103/PhysRevD.31.1841 [CrossRef] [PubMed]
19. Burdet, G.; Duval, C.; Perrin, M. Time Dependent Quantum Systems and Chronoprojective Geometry. *Lett. Math. Phys.* **1985**, *10*, 255–262. [CrossRef]
20. Duval, C.; Gibbons, G.W.; Horvathy, P. Celestial mechanics, conformal structures and gravitational waves. *Phys. Rev. D* **1991**, *43*, 3907–3922. [CrossRef]
21. Cariglia, M.; Duval, C.; Gibbons, G.W.; Horvathy, P.A. Eisenhart lifts and symmetries of time-dependent systems. *Ann. Phys.* **2016**, *373*, 631–654. [CrossRef]
22. Arnold, V.I. *Supplementary Chapters to the Theory of Ordinary Differential Equations*; Nauka: Moscow, Russia, 1978.
23. Arnold, V.I. *Geometrical Methods in the Theory of Ordinary Differential Equations*; Springer: New York, NY, USA, 1983. (In English)
24. Maslov, V.P.; Bouslaev, V.C.; Arnol'd, V.I. *Théorie des Perturbations et Méthodes Asymptotiques*; Dunod: Paris, France, 1972.
25. Arnold, V.I. Characteristic class entering in quantization conditions. *Funktsional'Nyi Anal. Ego Prilozheniya* **1967**, *1*, 1–14. [CrossRef]
26. Souriau, J.M. Construction explicite de l'indice de Maslov. Applications. *Lect. Notes Phys.* **1976**, *50*, 117–148. [CrossRef]
27. Burdet, G.; Perrin, M.; Perroud, M. Generating functions for the affine symplectic group. *Comm. Math. Phys.* **1978**, *58*, 241–254. [CrossRef]
28. Rezende, J. Quantum Systems with Time Dependent Harmonic Part and the Morse Index. *J. Math. Phys.* **1984**, *25*, 32643269, [CrossRef]
29. Aldaya, V.; Cossío, F.; Guerrero, J.; López-Ruiz, F.F. The quantum Arnold transformation. *J. Phys. A* **2011**, *44*, 065302. [CrossRef]
30. Guerrero, J.; López-Ruiz, F.F.; Aldaya, V.; Cossío, F.C. Symmetries of the quantum damped harmonic oscillator. *J. Phys. A* **2012**, *45*, 475303. [CrossRef]
31. Guerrero, J.; Aldaya, V.; López-Ruiz, F.F.; Cossío, F. Unfolding the quantum Arnold transformation. *Int. J. Geom. Meth. Mod. Phys.* **2012**, *9*, 1260011. [CrossRef]

32. Ermakov, V.P. Second order differential equations. Conditions of complete integrability. *Univ. Izv. Kiev Series III* **1880**, *9*, 1. [CrossRef]

33. Milne, W.E. The numerical determination of characteristic numbers. *Phys. Rev.* **1930**, *35*, 863. [CrossRef]

34. Pinney, E. The nonlinear differential equation $y'' + p(x)y + \frac{c}{y^3} = 0$. *Proc. Am. Math. Soc.* **1959**, *1*, 68.

35. Galajinsky, A. Geometry of the isotropic oscillator driven by the conformal mode. *Eur. Phys. J. C* **2018**, *78*, 72. [CrossRef]

36. Cariglia, M.; Galajinsky, A.; Gibbons, G.W.; Horvathy, P.A. Cosmological aspects of the Eisenhart–Duval lift. *Eur. Phys. J. C* **2018**, *78*, 314. [CrossRef]

37. Cheng, B. Exact propagator for the harmonic oscillator with time dependent mass. *Phys. Lett. A* **1985**, *113*, 293. [CrossRef]

38. Ilderton, A. Screw-symmetric gravitational waves: A double copy of the vortex. *Phys. Lett. B* **2018**, *782*, 22–27. [CrossRef]

39. Zhang, P.M.; Cariglia, M.; Elbistan, M.; Horvathy, P.A. Scaling and conformal symmetries for plane gravitational waves. *J. Math. Phys.* **2020**, *61*, 022502. [CrossRef]

40. López-Ruiz, F.F.; Guerrero, J. Generalizations of the Ermakov system through the Quantum Arnold Transformation. *J. Phys. Conf. Ser.* **2014**, *538*, 012015. [CrossRef]

41. Duval, C.; Horváthy, P.A.; Palla, L. Conformal properties of Chern-Simons vortices in external fields. *Phys. Rev.* **1994**, *D50*, 6658. [CrossRef] [PubMed]

42. Gibbons, G.W. Dark Energy and the Schwarzian Derivative. *arXiv* **2014**, arXiv:1403.5431.

43. Weisstein, E.W. Mathieu Function. 2003. Available online: https://mathworld.wolfram.com/MathieuFunction.html (accessed on 1 August 2021).

44. Cheng, B. Exact propagator for the one-dimensional time-dependent quadratic Lagrangian. *Lett. Math. Phys.* **1987**, *14*, 7–13. [CrossRef]

45. Jackiw, R. Introducing scale symmetry. *Phys. Today* **1972**, *25N1*, 23–27. [CrossRef]

46. Niederer, U. The maximal kinematical invariance group of the free Schrodinger equation. *Helv. Phys. Acta* **1972**, *45*, 802–810. [CrossRef]

47. Hagen, C.R. Scale and conformal transformations in galilean-covariant field theory. *Phys. Rev. D* **1972**, *5*, 377–388. [CrossRef]

48. De Alfaro, V.; Fubini, S.; Furlan, G. Conformal Invariance in Quantum Mechanics. *Nuovo Cim. A* **1976**, *34*, 569. [CrossRef]

49. Jackiw, R. Dynamical Symmetry of the Magnetic Vortex. *Ann. Phys.* **1990**, *201*, 83–116. [CrossRef]

50. Jackiw, R. Dynamical Symmetry of the Magnetic Monopole. *Ann. Phys.* **1980**, *129*, 183. [CrossRef]

51. Andrzejewski, K.; Prencel, S. Memory effect, conformal symmetry and gravitational plane waves. *Phys. Lett. B* **2018**, *782*, 421–426. [CrossRef]

52. Andrzejewski, K.; Prencel, S. Niederer's transformation, time-dependent oscillators and polarized gravitational waves. *Class. Quantum Gravity.* **2019**, *36*, 155008. [CrossRef]

53. Inzunza, L.; Plyushchay, M.S.; Wipf, A. Conformal bridge between asymptotic freedom and confinement. *Phys. Rev. D* **2020**, *101*, 105019. [CrossRef]

54. Dhasmana, S.; Sen, A.; Silagadze, Z.K. Equivalence of a harmonic oscillator to a free particle and Eisenhart lift. *arXiv* **2021**, arXiv:2106.09523.

55. Guha, P.; Garai, S. Integrable modulation, curl forces and parametric Kapitza equation with trapping and escaping. *arXiv* **2021**, arXiv:2104.06319.

 *symmetry*

Article

What Is the Size and Shape of a Wave Packet?

Larry S. Schulman

Physics Department, Clarkson University, Potsdam, NY 13699-5820, USA; lschulma@clarkson.edu

Abstract: Under pure quantum evolution, for a wave packet that diffuses (like a Gaussian), scattering can cause localization. Other forms of the wave function, spreading more rapidly than a Gaussian, are unlikely to localize.

Keywords: spreading wave function; scattering; Localization

Citation: Schulman, L.S. What Is the Size and Shape of a Wave Packet?. *Symmetry* **2021**, *13*, 527. https://doi.org/10.3390/sym13040527

Academic Editor: Georg Junker

Received: 17 February 2021
Accepted: 16 March 2021
Published: 24 March 2021

Publisher's Note: MDPI stays neutral with regard to jurisdictional claims in published maps and institutional affiliations.

What is the size and shape of a wave packet? I am talking about a wave packet of a particle (atom or molecule) in a gas. Is it a plane wave that fills the container? Or is it a microscopic (perhaps $\leq 100\,\text{nm}$) object? I am not talking about a situation where there is a potential holding the particle in some region, like a hydrogen atom. The only things around are other particles, the same or different. (Please see Appendix A). Strangely, this subject has not attracted a lot of attention.

Our conclusion considers a number of possibilities. Should the eventual wave functions be Gaussian or Gaussian-like (to be later defined) then, yes, there is localization. If not, probably no. However, we give an argument that the wave function most likely does become Gaussian. (Generally speaking, people use Gaussians in descriptions, but this is not always warranted.)

This paper has three parts: arguments for a Gaussian, localization in that case and the case(s) that it is not a Gaussian. We emphasize that the first part (eventually) treats non-Gaussians that become Gaussian after a number of collisions.

Arguments for a Gaussian. The wave function for a Gaussian would be

$$\psi(r) = \frac{1}{(\pi\sigma^2)^{3/4}} \exp\left(-\frac{(r-r_0)^2}{2\sigma^2} + \frac{i}{\hbar}p_0(r+r_0)\right). \tag{1}$$

In one dimension, a Gaussian in momentum space is

$$\phi(p) = \left(\frac{\sigma^2}{\pi\hbar^2}\right)^{1/4} \exp\left(-\frac{\sigma^2(p-p_0)^2}{2\hbar^2} - \frac{i}{\hbar}px_0\right). \tag{2}$$

(The following paragraph is the treatment in [1]; for more details, see that reference.) We suppose there is a particle of a different mass and that they scatter. The conservation laws (once they are far enough apart that they do not interact) are

$$P_{\text{final}} = P_{\text{initial}} \text{ and } p_{\text{final}} = -p_{\text{initial}}, \tag{3}$$

where P and p are collective coordinates: $P = p_1 + p_2$ and $p = (m_2p_1 - m_1p_2)/M$. In these relations, p_j and m_j ($j = 1, 2$) are the momenta and masses of each of the particles, $\frac{1}{\mu} = \frac{1}{m_1} + \frac{1}{m_2}$ and $M = m_1 + m_2$. Furthermore, we let $\mu_i = \frac{m_i}{M}$ ($i = 1, 2$). It turns out that, for calculation of the spread, all that matters is the real part of the ($2\times$ negative) exponent. This changes from $[\sigma_1^2 p_1^2 + \sigma_2^2 p_2^2] = [\sigma_1^2(\mu_1 P + p) + \sigma_2^2(\mu_2 P - p)]$, following Equation (3), to

$$Q(p_1, p_2) \equiv p_1^2[\sigma_1^2(\mu_1 - \mu_2)^2 + 4\mu_2^2\sigma_2^2] + p_2^2[\sigma_2^2(\mu_2 - \mu_1)^2 + 4\mu_1^2\sigma_1^2]$$
$$+ 4(\mu_1 - \mu_2)p_1p_2(\mu_1\sigma_1^2 - \mu_2\sigma_2^2), \tag{4}$$

As shown in [1], provided μ_1 and μ_2 are significantly different from one another, within a few scattering events the "cross" term in $p_1 p_2$ vanishes. This means that the wave functions have ratio of spread such that $m\sigma^2$ is constant and there is no cross term—nothing to decohere (and the von Neumann entropy is maximum). In three dimensions the same thing happens, but is more difficult to show [2–4]. That, however, is not our main point.

What happens in one dimension if the initial functions are not a Gaussian? We suppose the wave function has the form $\exp(-\sum_2^\infty a_n x^n)$. This form can fit various other functional forms, e.g., $1/(1+x^2)^2$ to within 0.06 using 4 coefficients (i.e., elements of the set $\{a_n\}$). (Please see Appendix B) . Following the earlier method (based on Equation (3)), we find that 4th order terms obey

$$
\begin{aligned}
a_4^{(1)'} &= a_4^{(1)}(\mu_1-\mu_2)^4 + a_4^{(2)}2^4\mu_2^4 \\
a_4^{(2)'} &= a_4^{(2)}(\mu_1-\mu_2)^4 + a_4^{(1)}2^4\mu_1^4 ,
\end{aligned}
\tag{5}
$$

with $a_4^{(j)}$ $(j=1,2)$ the coefficients of x^4 for the respective wave functions. In general, if the deviation from a Gaussian begins with a term p_k^{2n} there is a matrix that takes one from the values of the coefficients multiplying these terms from before scattering to after scattering. That matrix is

$$
\begin{pmatrix}
(\mu_1-\mu_2)^{2n} & (2\mu_2)^{2n} \\
(2\mu_1)^{2n} & (\mu_1-\mu_2)^{2n}
\end{pmatrix} .
\tag{6}
$$

The eigenvalues of this matrix are (for all $n>1$) below 1 for $m_1 \neq m_2$ and $m_j \neq 0$ $(j=1,2)$.

What happens if there is no other-mass particle? I do not know. I would not have thought it should make a difference, but I do not have a proof. In practice, there is almost always some impurity, but it may scatter rarely.

Thus, there is an indication that in one dimension the wave function approaches a Gaussian. In higher dimension—in particular 3—I do not have definitive results. It is still true [1] that for Gaussians the spread approaches a maximum of von Neumann entropy and (if there are two types of particles, #1 and #2 then) $m_1\sigma_1^2 = m_2\sigma_2^2$.

For three dimensions, there are many ways that n^{th} power terms can occur; for #1 one can have anything of the form $p_{1x}^{n_1} p_{1y}^{n_2} p_{1z}^{n_3}$ with $n_1+n_2+n_3 = n$, and similarly for #2, leading to $(n+2)(n+1)$ coefficients for the two of them. (Please see Appendix C). Moreover, as discussed in [1], the post-scattering values of p_1 and p_2 involve a rotation, $R \in \mathrm{SO}(3)$, not just a flip. Thus, $p_1' = (\mu_1 I + \mu_2 R)p_1 + \mu_1(I-R)p_2$ (and $1 \leftrightarrow 2$ for p_2').

I have examined cubic and quartic components of the logarithm of the wave function. All the indicated operations have been carried out, the cross terms involving momenta of #1 and of #2 have been dropped, and a rather complex recursion for the plethora of coefficients evaluated numerically. Sample results are shown in Figures 1 and 2. The "time" represents the number of scattering events, and the ordinate is the sum total of the absolute values of all cubic or quartic coefficients. Remarkably, they all tend to zero.

Evaluations beyond quartic represent a further problem in symbolic manipulation, but based on the higher-power evidence of one dimension together with the cubic and quartic evidence in three dimensions, it is reasonable to make the claim that all higher power coefficients tend to zero under decoherence.

Is this a proof that eventually everything decoheres to a Gaussian? Absolutely not, but it is an indication.

Localization with Gaussian-like behavior. Scattering can localize. This may be surprising, since some may hold that a wave function can only spread. It turns out (as shown in [2]) that scattering can act like a measurement, that scattering alone can localize. Here we extend that result.

I do not deal with the effects of temperature (cf. [5]), nor with off-diagonal elements of the density matrix [6,7]. I am concerned with pure quantum behavior. Nor do the conclusions depend on interpretation, i.e., they are independent of whether one subscribes to the Copenhagen interpretation (in its many variants), Many Worlds, or some other theory.

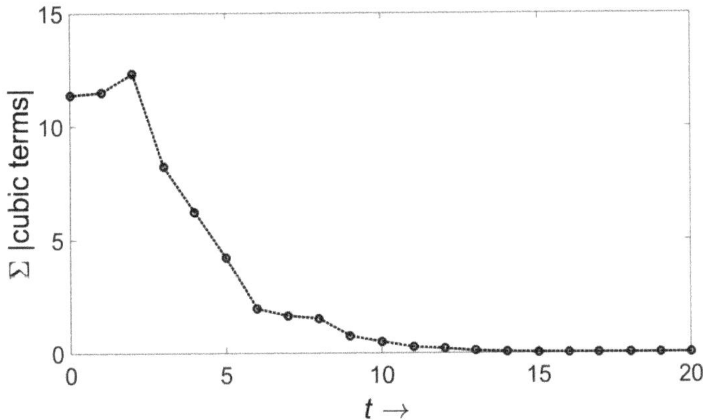

Figure 1. Reduction of non-Gaussian terms with unequal mass scattering in three dimensions. Because that the problem has an intrinsic nonlinearity, the initial coefficients were varied. Usually they were taken as random between 0 and 1, but on some occasions they were taken to be 20 times that. In all cases there was convergence to zero. The programs to establish this were combinations of symbolic manipulation and numerical evaluation.

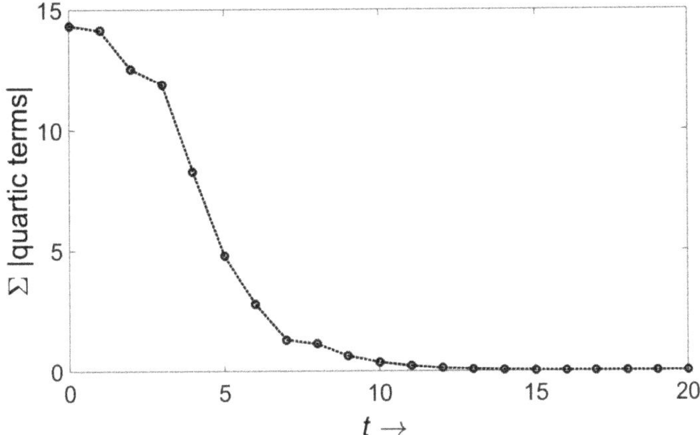

Figure 2. As in Figure 1, except that this looks at quartic terms. As in the other example, all terms shrink to zero.

I will briefly review the results of [2] and then turn to the extension. (There is a slight change in notation: instead of σ^2, we use $2\Delta^2$ for convenience in matching results.) The principal consequence of [2] is that, assuming the wave function is a Gaussian, particles do not spread indefinitely. The proof in [2] is a self-consistency argument. We assume two normalized Gaussians in 3 dimensions of the form

$$\psi(r_1, r_2, 0) = \frac{1}{(2\pi\Delta^2)^{3/2}} \exp\left(-\frac{(r_1 + r_0)^2}{4\Delta^2} + \frac{i}{\hbar}p_1(r_1 + r_0) - \frac{(r_2 - r_0)^2}{4\Delta^2} + \frac{i}{\hbar}p_2(r_2 - r_0) \right). \tag{7}$$

with parameters r_0, p_1, p_2 and Δ. We assume $p_1 \approx -p_2$, so that at time $m|r/p_1|$ they scatter. Assuming no interaction, at time t this becomes

$$\psi(r_1, r_2, t) = \left(\frac{\Delta^4 + (\frac{\hbar t}{2m})^2}{2\pi\Delta^2} \right)^{-3/2} \exp\left(-\frac{(r_1 + r_0 - \frac{p_1 t}{m})^2}{4\left(\Delta^2 + i\frac{\hbar t}{2m}\right)} + \frac{i}{\hbar} p_1(r_1 + r_0) - \frac{i}{\hbar}\frac{p_1^2 t}{2m} \right)$$

$$\times \exp\left(-\frac{(r_2 - r_0 - \frac{p_2 t}{m})^2}{4\left(\Delta^2 + i\frac{\hbar t}{2m}\right)} + \frac{i}{\hbar} p_2(r_2 - r_0) - \frac{i}{\hbar}\frac{p_2^2 t}{2m} \right). \tag{8}$$

using center of mass coordinates, the exponent is

$$-\frac{(R + \frac{Pt}{2m})^2}{2\left(\Delta^2 + i\frac{\hbar t}{2m}\right)} - \frac{(r + 2r_0 + \frac{pt}{m/2})^2}{2\left(4\Delta^2 + i\frac{\hbar t}{m/2}\right)} - \frac{i}{\hbar}(RP + (r - 2r_0)p) - \frac{i}{\hbar}\left(\frac{P^2}{4m} + \frac{p^2}{m} \right)t, \tag{9}$$

where $R = \frac{1}{2}(r_1 + r_2)$, $r = r_1 - r_2$, and correspondingly for P and p. Now set $t \approx m|r/p_1|$ (or just afterward). They scatter, but nothing happens to P. In the center of mass, we use the Born approximation. Up to normalization, the new wave function is $\delta t \frac{i}{\hbar} V(r)\Psi(R, r, t)$, where we also assumed the interaction was brief, taking place during a short time δt. With a further condition that $V(r) = V_0 \exp\left(-\frac{r^2}{4a^2}\right)$, for convenience in integrating, we obtain

$$|\Psi|^2 = \exp\left(-\frac{r^2}{2a^2} - \frac{(R + \frac{Pt}{2m})^2}{\left|\Delta^2 + i\frac{\hbar t}{2m}\right|} - \frac{(r + 2r_0 + \frac{pt}{m/2})^2}{\left|4\Delta^2 + i\frac{\hbar t}{m/2}\right|} \right). \tag{10}$$

(Normalization cancels and can be ignored.) Using this wave function, one can calculate the spread in r (Δr), in R (ΔR) and from them Δr_1 and Δr_2. The latter quantities are then set equal to the original Δr_1 and Δr_2 (both equal to Δ). Using as the time between collisions $\frac{\text{scattering length}}{\text{velocity}}$ (and mass values) we find that indeed the original Δ can be set equal to the final values of spread. To make equations simpler, we define

$$\theta \equiv \frac{\hbar t}{m} = \frac{\hbar \ell}{mv}, \tag{11}$$

where ℓ is the scattering length and v a typical velocity. For any given gas, the range of θ is fixed.

The essential steps in the foregoing derivation are the estimation of Δr, R and ΔR. It is these that can be generalized.

The first step is to formalize V. Instead of an exponential of range a, we take a potential that is cut off at distance a. This restricts the location of r to be within a distance a of R. In other words, $\Delta r \leq \sqrt{r \cdot r} = 3a^2$. This is an additional assumption and may require a larger "a" than was previously posited.

The expectation of R is simply $\langle R \rangle = \frac{Pt}{2m}$, since it is unaffected by the interaction/scattering.

The *spread* in R is another matter and is the principal source of uncertainty in our calculation. For a Gaussian-like initial wave function, we can give estimates. From our previous work [1], we have

$$\langle (\Delta R)^2 \rangle = \langle R^2 \rangle - \langle R \rangle^2 = 3\frac{4\Delta^4 + \theta^2}{8\Delta^2}. \tag{12}$$

Now we would like to weaken the assumptions. It turns out that the most important property (for our purposes) of the Gaussian is its diffusive behavior. This assumption about the spread in the center of mass coordinate is weakened in a specific way: in place of the 8 that appears in the denominator, we allow smaller values, γ. (In this sense, the wave function is Gaussian-like.) Specifically, $4 < \gamma \leq 8$ and

$$\langle (\Delta R)^2 \rangle = 3\,\frac{4\Delta^4 + \theta^2}{\gamma \Delta^2}\,.$$

(13)

We now use $\langle \Delta r_1^2 \rangle = \langle \Delta r_2^2 \rangle = \langle \Delta R^2 \rangle + \frac{1}{4}\langle \Delta r^2 \rangle$ to arrive at the denominators for the new wave functions of r_1 and r_2:

$$\langle (\Delta r_{new})^2 \rangle = \langle (\Delta r_k)^2 \rangle = 3a^2 + 3\,\frac{4\Delta^4 + \theta^2}{\gamma \Delta^2}\,, \qquad k = 1.2\,.$$

(14)

Setting the old Δ equal to the new one, we arrive at a self-consistency criterion for the spread

$$3\Delta^2 = 3a^2 + 3\,\frac{4\Delta^4 + \theta^2}{\gamma \Delta^2}\,.$$

(15)

This is a quadratic equation whose solution is

$$\Delta^2 = \frac{a^2 \gamma}{4(\gamma - 4)} \pm \sqrt{\frac{\theta^2}{\gamma - 4} + \left(\frac{a^2 \gamma}{4(\gamma - 4)}\right)^2}\,.$$

(16)

Taking $\ell = 70\,\mathrm{nm}$, $v = 500\,\mathrm{m/s}$, $a = 1\,\mathrm{nm}$ and $m = 29\,\mathrm{gm}/6 \times 10^{23} \approx 5 \times 10^{-23}\,\mathrm{gm}$ and $\gamma = 4.1$ gives a value of 4.5 nm, much less than the mean free path. (Please see Appendix D). These are typical parameters for the air and the wave function is localized. Note though that (for air) there is overlap: (number density)$^{1/3} \approx (0.025 \times 10^{-27})^{1/3} \approx 3\,\mathrm{\AA}$. Therefore, although the wave functions occupy common volumes, they mostly do not interact.

Not a Gaussian. The essential feature of non-Gaussian wave functions is that the spread in R grows. I have not examined the "boundary," that is, the form of, or parameters in, wave functions that eventually become Gaussian and those that do not. I will though examine various wave functions that are non-Gaussian. Consider, for example, an exponential

$$\psi(r, 0) = \frac{\lambda^3}{8\pi}\exp(-\lambda r)\,.$$

(17)

We assume that both scattering particles have this form, with the same value of λ. The spread in the relative coordinate Δr is still bounded by a (or $3a$) but the center of mass coordinate grows. The spread for each, before collision, is $1/\lambda$. The center of mass coordinate for equal mass particles is $R = (r_1 + r_2)/2$, so that the spread for of the center of mass coordinate is also $1/\lambda$. As a result, the center of mass can also be taken of the form Equation (17). Now apply the free propagator numerically and fit the result to an exponential. Except for particular values of λ the value of the spread has *increased*. This means there can be no self-consistent solution (as there was for a Gaussian).

The same happens for the wave function taken as a power.

Conclusions. The point of this is not that no wave function spreads. Rather, it places bounds on spreading for certain wave functions and makes it plausible that scattering is sometimes like a measurement, pinning a particle to a small, localized region.

Funding: This research received no external funding.

Institutional Review Board Statement: Not applicable.

Informed Consent Statement: Not applicable.

Data Availability Statement: Not applicable.

Conflicts of Interest: The authors declare no conflict of interest.

Appendix A

In particular, we do not deal with chains of oscillators or single oscillators, which according to [8,9] (and other literature) become, by decoherence, coherent states, i.e., Gaussian.

Appendix B

The function $1/(1 + x^2)^2$ is not symmetric about $x = 0$; hence, it requires odd terms in the expansion $\exp(-\sum_2^\infty a_n x^n)$. This is accomplished with coefficients a_3 and a_5 that are 10^{-8} or less. The fourth order term is about $1/100$ of the quadratic term (which is about 1.46).

Appendix C

This is the number of possibilities for both particles and is calculated by imagining the interval $[0, n]$ as having partitions at 2 (integer) locations (and multiplying by 2). Just to be clear, the 3-dimensional tendency to have a Gaussian wave function is new material.

Appendix D

The minus sign in Equation (16) is spurious and gives an imaginary value for $\sqrt{\Delta^2}$.

References

1. Schulman, L.S. Evolution of wave-packet spread under sequential scattering of particles of unequal mass. *Phys. Rev. Lett.* **2004**, *92*, 210404. [CrossRef] [PubMed]
2. Gaveau, B.; Schulman, L.S. Reconciling Kinetic and Quantum Theory. *Found. Phys.* **2020**, *50*, 55–60. [CrossRef]
3. Schulman, L.S.; Schulman, L.J. Convergence of matrices under random conjugation: wave packet scattering without kinematic entanglement. *J. Phys. A* **2006**, *39*, 1717–1728. [CrossRef]
4. Schulman, L.S.; Schulman, L.J. Wave packet scattering without kinematic entanglement: Convergence of expectation values. *IEEE Trans. Nanotech.* **2004**, *4*, 8–13. [CrossRef]
5. Ford, G.W.; O'Connell, R.F. Wave packet spreading: Temperature and squeezing ef- fects with applications to quantum measurement and decoherence. *Am. J. Phys.* **2002**, *70*, 319–324. [CrossRef]
6. Joos, E.; Zeh, H. The Emergence of Classical Properties Through Interaction with the Environment. *Z. Phys. B* **1985**, *59*, 223–243. [CrossRef]
7. Gaveau, B.; Schulman, L.S. Decoherence, the Density Matrix, the Thermal State and the Classical World. *J. Stat. Phys.* **2017**, *169*, 889–901. [CrossRef]
8. Tegmark, M.; Shapiro, H.S. Decoherence produces coherent states: An explicit proof for harmonic chains. *Phys. Rev. E* **1994**, *50*, 2538–2547. [CrossRef] [PubMed]
9. Zurek, W.H.; Habib, S.; Paz, J.P. Coherent States via Decoherence. *Phys. Rev. Lett.* **1993**, *70*, 1187–1190. [CrossRef] [PubMed]

MDPI

St. Alban-Anlage 66

4052 Basel

Switzerland

Tel. +41 61 683 77 34

Fax +41 61 302 89 18

www.mdpi.com

Symmetry Editorial Office

E-mail: symmetry@mdpi.com

www.mdpi.com/journal/symmetry